AF522241

Inland Water Resources
India

Inland Water Resources India

Volume 1

Editors
M.K. DURGA PRASAD
&
P. SANKARA PITCHAIAH
Dept. of Geology
Nagarjuna University
Nagarjunanagar-522 510
Andhra Pradesh

1999
Discovery Publishing House
New Delhi-110 002

First Published-1999

ISBN 81-7141-511-3 (Set)

Published by :
DISCOVERY PUBLISHING HOUSE
4831/24, Ansari Road, Prahlad Street,
Darya Ganj, New Delhi-110 002 (*INDIA*)
Phone: 3279245
Fax: 91-11-3253475

Laser Typeset at:
BHARGAVA PRINTO
1643, Dakhni Rai Street,
Darya Ganj, Delhi-2

Printed at:
Arora Offset Press,
Delhi-110092.

PREFACE

In India, the vast Inland Water Resources are ill managed and improperly utilised. As such, the sustainable clean waters and their resources are becoming rare commodities! It is a fact that the optimal and sustainable utilisation of these waters and their resources require scientific and technical input from various disciplines. Hence, an attempt is made to gear up the valuable scientific information collected by the scientific personnel from all over India belonging to various disciplines such as Agriculture, Aquaculture, Botany, Chemistry, Civil Engineering, Fisheries, Geography, Geology, Horticulture, Physics, Remote Sensing, Toxicology and Limnology. We hope the cluster of knowledge presented in these volumes will be useful to the reader.

M.K. Durga Prasad
P. Sankara Pitchaiah

Contents

Section I

Water Quality and Hydrochemistry

Section II

Biotic Components of Water

List of Contributors and Addresses

1. V.K. Saxena
   NGRI, Uppal Road, Hyderabad-500 007.
2. Yamuna Singh
   Atomic Minerals Division, Begumpet, Hyderabad-500 016.
3. S. Sathyamoorthy
   Dept. of Geology, National College, Tiruchirapalli-600 001.
4. N. Dhakshinamoorthy
   Dept. of Met. Engg., Regional Engineering College, Tiruchirapalli-600 015.
5. P.R. Srinivasan
   Dept. of Geology, National College, Tiruchirapalli-600 001.
6. A. Balasubramanian
   Dept. of Geology, VOC College, Tuticorin-628 008.
7. A. Ramanathan
   Dept. of Earth Sciences, IIT, Bombay-400 076.
8. D. Chandrasekhram
   Dept. of Earth Sciences, IIT, Bombay-400 076.
9. N. Someswara Rao
   School of Chemistry, Andhra University, Visakhapatnam-530 003.
10. T.N.V. Venkateswara Rao
    School of Chemistry, Andhra University, Visakhapatnam-530 003.
11. D. Jagannadha Rao
    School of Chemistry, Andhra University, Visakhapatnam-530 003.
12. B. Narasimha Murthy
    School of Chemistry, Andhra University, Visakhapatnam-530 003.
13. D. Srinivasu
    Dept. of Geology, Nagarjuna University, Nagarjuna Nagar-522 510.
14. P. Sankara Pitchaiah
    Dept. of Geology, Nagarjuna University, Nagarjuna Nagar-522 510.
15. T.N. Tiwari
    Dept. of Physics, Regional Engineering College, Rourkela-769 008.

16. M. Priyadarshni
Dept. of Physics, Regional Engineering College, Rourkela-769 008.
17. P.R. Subramanian
Central Ground Water Board, Nagpur-440 010.
18. N. Kittu
Director, Central Ground Water Board, Nagpur-440 010.
19. A.K. Pandit
Center of Research Development, The Univ. of Kashmir, Srinagar-190 006.
20. A. Sayeswari
Dept. of Botany, Kakatiya University, Warangal-506 009.
21. M. Singara Charya
Dept. of Botany, Kakatiya University, Warangal-506 009.
22. K.S. Tilak
Dept. of Zoology, Nagarjuna University, Nagarjuna Nagar 522 510.
23. A. Annamani
Dept. of Zoology, Nagarjuna University, Nagarjuna Nagar 522 510.
24. S. Jhansi Lakshmi
Dept. of Zoology, Nagarjuna University, Nagarjuna Nagar 522 510.
25. T. Anita Susan
Dept. of Zoology, Nagarjuna University, Nagarjuna Nagar 522 510.
26. S.M. Anwar Batcha
Dept. of Zoology, Jamal Mohamed College, Tiruchirapalli-600 020.
27. P. Padmavati
Centre for Aquaculture Research Training and Education, Nagarjuna University, Nagarjuna Nagar-522 510.
28. M.K. Durga Prasad
Dept. of Zoology, Nagarjuna University, Nagarjuna Nagar-522 510.
29. N. Jeeji Bai
Shri AMM Murugappa Chettiar Research Centre, Tharamani, Madras-600 113.
30. D. Lakshmi
Sri Sarada College, Salem-636 016.
31. Syed Hasib Ahmad
Dept. of Fisheries, Bihar Veterinary College, Patna-800 014.
32. A.K. Singh
Freshwater Biology Laboratory, B.D.E. College, Patna-800 001.

33. K.S. Rao
School of Studies in Zoology, Vikram University, Ujjain-456 010.
34. S. Shrivastava
School of Studies in Zoology, Vikram University, Ujjain-456 010.
35. U. Choubey
School of Studies in Zoology, Vikram University, Ujjain-456 010.
36. N.K. Dad
School of Studies in Zoology, Vikram University, Ujjain-456 010.
37. Anjana Dewanji
Biological Science Division, Indian Statistical Institute 203, B.T. Road, Calcutta-700 035.
38. Gunwant Joshi
M.P. Pollution Control Board, Bhopal-462 016.
39. A.D. Adoni
Dept. of Botany, Dr. Harisingh Gour Viswavidyalaya, Sagar-470 003.
40. S.G. Sarwar
C/O Sved Lodge, Boulevard, Srinagar-190 001.
41. Sunanda Chanda
Leaf Protein Research Unit, Indian Statistical Institute 203, Barrackpore Trunk Road, Calcutta-700 035.
42. P. Ajitha
Centre for Aquaculture Research Training and Education, Nagarjuna University, Nagarjuna Nagar-522 510.
43. P.S.R. Anjaneyulu
Centre for Aquaculture Research Training and Education, Nagarjuna University, Nagarjuna Nagar-522 510.
44. S. Vijaya Mohan
Gandhi Medical College, Hyderabad-500 029.
45. V. Radhakrishan
Dept. of Geology, VOC College, Tuticorin-628 008.
46. R. Tirugnana Sambandam
Dept. of Geology, VOC College, Tuticorin-628 008.
47. R. Chellasamy
Dept. of Geology, VOC College, Tuticorin-628 008.
48. Archana Sarma
Dept. of Bioscience, Ravishankar University, Raipur-492 010.
49. M.L. Naik
Dept. of Bioscience, Ravishankar University, Raipur-492 010.

50. D.R.R. Sharma
School of Chemistry, Andhra University, Visakhapatnam-530 003.
51. K.G. Naidu
School of Chemistry, Andhra University, Visakhapatnam-530 003.
52. M. Venkata Rao
School of Chemistry, Andhra University, Visakhapatnam-530 003.
53. Sanjay Sharma
23, Vikas Nagar, Gwalior-474 002.
54. R. Mathur
School of Studies in Zoology, Jiwaji University, Gwalior-474 011.
55. Asha Mathur
Dept. of Zoology, Kamla Raja Girls Post-Graduate College, Gwalior-474 009.
56. A.K. Rastogi
Dept. of Civil Engineering, IIT, Bombay-400 076.
57. S. Mohan
Dept. of Civil Engineering, IIT, Madras-600 036.
58. H. Raman
Dept. of Civil Engineering, IIT, Madras-600 036.
59. K. Srinivasa Raju
Dept. of Civil Engineering, IIT, Madras-600 036.
60. K. Narasimha Murthy
WALAMTARI, I & CAD Department, Himayatsagar, Hyderabad-500 030.
61. P.K. Das
C/o H.K. Sharma, 32A/14 Suren Sarkar Road, Calcutta-700 010.
62. J.C.V. Sastry
Dept. of Geology, University of Mysore, Mysore-570 006.
63. Ashok Kumar Saini
Dept. of Soil and Water Engineering, PAU, Ludhiana-141 004.
64. M.P. Kaushal
Dept. of Soil and Water Engineering, PAU, Ludhiana-141 004.
65. S.R. Singh
WTC, Eastern Region, Bhubaneswar.
66. S.K. Gupta
Central Soil Salinity Research Institute, Karnal-132 001.

67. A.S. Rayamane
Dept. of Geography, Karnataka University, Dharwad-3.

68. S.S. Naregal
Dept. of Geography, Karnataka University, Dharwad-3.

69. N. Thangarajan
NGRI, Uppal Road, Hyderabad-500 007.

70. V.S. Singh
NGRI, Uppal Road, Hyderabad-500 007.

71. K. Subramaniam
NGRI, Uppal Road, Hyderabad-500 007.

72. S.M. Ramaswamy
School of Earth Sciences, Bharathidasan University, Tiruchirapalli-620 023.

73. V. Venkatasubramanian
School of Earth Sciences, Bharatidasan University, Tiruchirapalli-620 023.

74. K. Joan Sam
Dept. of Environmental Sciences, Bishop Heber College, Tiruchirapalli-620 017.

75. G. Edwin Chandrasekhar
Dept. of Environmental Sciences, Bishop Heber College, Tiruchirapalli-620 017.

1

Saltwater Intrusion Mechanism along the Coastal Aquifer, Malappuram District, Kerala

D. CHANDRASEKHARAM

Introduction

Sea-water intrusion in unconfined aquifers is a common problem in the coastal areas.

Several mechanisms, such as the Ghyben-Herzberg and negative gradient mechanisms cause the sea water to enter the aquifers. Besides the unconfined aquifers, confined aquifers also show sea water mixing as exemplified by the thermal springs along the coast of Maharashtra. Here, the meteoric water penetrating down to deeper levels gets heated due to the natural thermal gradient and emerge on to the surface through fractures as thermal springs. These thermal springs also contain seawater component in them. These springs, during their descent or ascent, are mixing with sea water at some depth (Chandrasekharam *et al.*, 1989).

In this paper, the sea water intrusion mechanism operation along the coastal belt of Malappuram district of Kerala is presented using chemical characteristics of groundwater and aquifer-sea water mixing chemical model.

Open dug wells tapping the top unconfined laterite aquifer provide water for drinking. The number of such wells vary from 200 to 300 in an area of about 1 sq.km. Coastal alluvium and laterite form the main aquifer. Some of the wells show evidence of sea water mixing. The data presented here has been collected for three years from 1980 to 1982 from six representative wells in an area of 10 sq.km. (Fig. 1.1).

Hydrogeology and Chemistry

The depth of the wells vary from 2 to 7m from the ground level. Four wells located near the coastline tap sand plus laterite aquifer whereas the remaining two wells located away from the coast tap the laterite aquifer. The porosity and permeability of sand and laterite of this region vary from 1426 to $2285*10^{-5}$ cm/sec respectively (Mallikarjuna *et al.*, 1979). The area receives an average annual rainfall of 3000 mm, the major share of which is generally received during June-August. Water levels

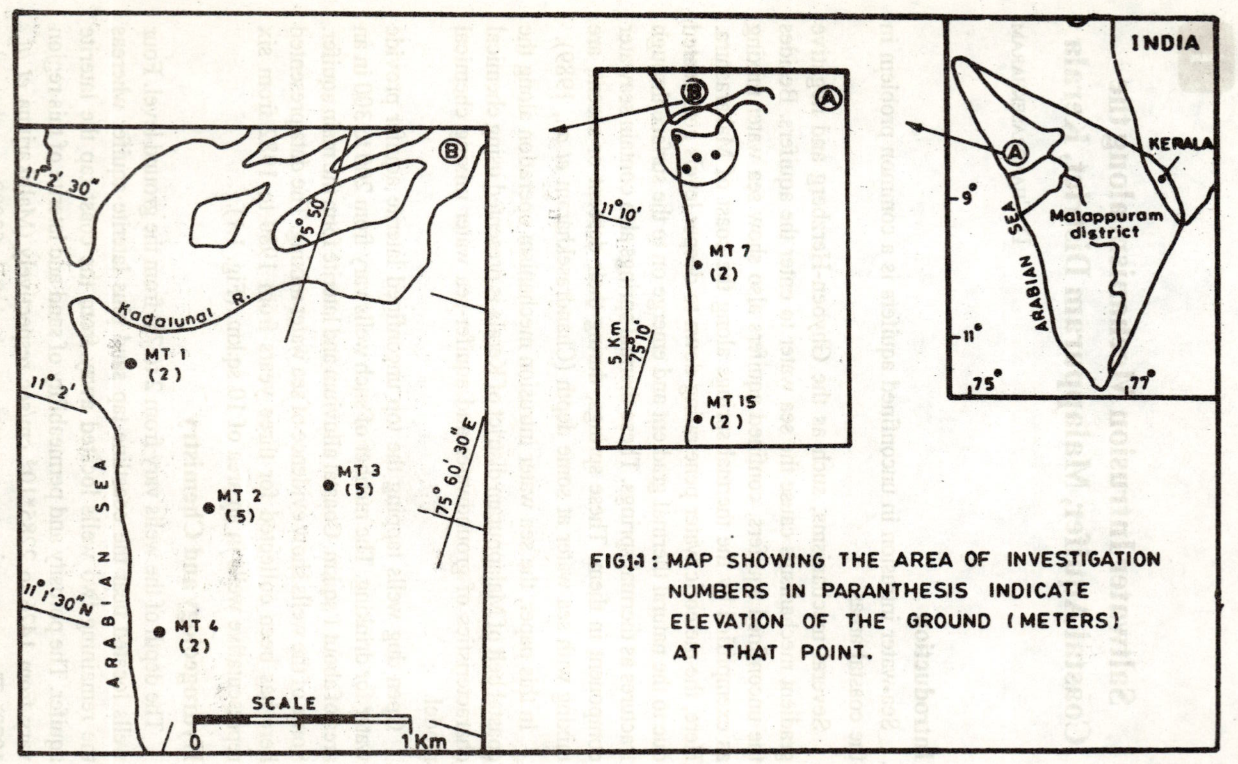

FIG 1 : MAP SHOWING THE AREA OF INVESTIGATION NUMBERS IN PARANTHESIS INDICATE ELEVATION OF THE GROUND (METERS) AT THAT POINT.

from the observation wells have been recorded during pre-monsoon (April to May), monsoon and after the cessation of rains (December to January). The water table as referred to the ground and mean sea levels is shown in Table 1.1. The ground elevation at the well sites is marked in Fig. 1.1. It can be seen from Table 1.1 that groundwater gradient generally tends towards the sea. However, at certain localities, (MT4 and MT2) the water table falls below the mean sea level giving rise to a negative gradient. In well MT4 the gradient is permanently reversed while in wells MT2 and MT15 the gradient reversal is seasonal. Considering the density of the open wells, this phenomena appears to be due to excess withdrawal of water through these wells.

Water samples have been collected at regular intervals specified above and analysed for the dissolved constituents adopting the standard methods specified by APHA (1975). pH and electrical conductivity have been measured using field pH and conductivity meter. Na and K have been measured using a flame photometer. The results are shown in Tables 1.2 and 1.3.

Results and Discussion

Water occurring in the aquifer should generally be of bicarbonate type. However, from the chemical data reported in Table 1.3 it is seen that chloride ion dominates over bicarbonate ion in certain wells. Since, the aquifer do not contain appreciable chloride bearing mineral the Cl^- ion is contributed by the sea through rain and through the aquifer (sea water intrusion). Thus Cl, SO_4 and HCO_3 ions can be used to classify the water into bicarbonate group and chloride group depending upon the relative abundance of these ions. Such distinction is clearly seen when the analyses are plotted on the Piper (1944) diagram (Fig. 1.2). The plots falling in the chloride facies show $Cl+SO_4/HCO_3$ ratio above unity and those with the above ratio below unity fall within the bicarbonate facies. The well waters with $Cl+SO_4/HCO_3$ above unity evidently show mixing of seawater. During monsoon, due to rise in water table level and increase in the groundwater flow rate, the plots are expected to shift to the bicarbonate field. On the contrary, these plots though show seasonal $Cl+SO_4/HCO_3$ variation, lie within the field of chloride facies indicating the presence of sea water component in it. Thus a constant mixing of sea water and aquifer water is taking place in the above wells. Exception to this are wells MT1, MT15 and MT2. In the case of well MT1 the water table has hydraulic continuity with the adjacent river and sea. Hence, the chemistry of water in this well is controlled by the surface water body and groundwater. The surface water mixes with sea water during

Table 1.1 a: Depth to water table from the ground level in the wells

Well No:		Premonsoon			Monsoon			Postmonsoon		
		1980	*1981*	*1982*	*1980*	*1981*	*1982*	*1980*	*1981*	*1982*
MT1	(2.0)	1.4	2.0	1.6	1.8	1.8	1.5	1.2	1.5	1.8
MT2	(7.1)	5.7	5.3	5.4	–	2.3	0.2	4.0	4.2	4.9
MT3	(4.3)	3.9	3.5	3.7	1.4	1.2	1.0	2.0	2.4	2.9
MT4	(3.2)	2.6	2.8	2.7	2.4	2.2	2.3	2.5	2.6	–
MT7	(4.2)	3.0	3.2	–	2.7	2.8	–	3.6	2.5	–
MT15	(2.4)	1.8	2.0	–	0.8	2.3	–	1.6	1.5	–

b: Water levels with reference to mean sea level

Well No:		Premonsoon 1980	Premonsoon 1981	Premonsoon 1982	Monsoon 1980	Monsoon 1981	Monsoon 1982	Postmonsoon 1980	Postmonsoon 1981	Postmonsoon 1982
MT1	+0.6	0.0	+0.4	+0.2	+0.2	+0.5	+0.8	0.5	+0.2	
MT2	–0.7	–0.3	–0.4	–	+2.7	+4.8	+1.0	+0.8	+0.1	
MT3	+1.2	+1.5	+1.3	+3.6	+3.8	+4.0	+3.0	+2.6	+2.1	
MT4	–0.6	–0.8	–0.7	–0.4	–0.2	–0.3	–0.5	–0.6	–0.7	
MT7	–1.0	–1.2	–	–0.7	–0.8	–	–1.6	–1.5	–	
MT15	+0.2	0	–	+0.1	–0.3	–	+1.4	+1.5	–	

All values in meters. Numbers in parenthesis indicate total depth of the well from ground level.

Table 1.2 : pH and Ec of water samples from the wells

	Premonsoon			Monsoon		Postmonsoon		Well No :
	1980	1981	1982	1980	1981	1980	1982	
pH	6.6	6.3	6.8	7.9	7.3	6.8	6.5	MT1
Ec	4400	4000	3600	350	515	695	665	
pH	6.9	6.9	6.2	7.6	7.6	6.8	7.8	MT2
Ec	327	312	170	110	184	300	228	
pH	5.7	5.4	6.0	7.0	6.7	6.5	7.7	MT3
Ec	114	134	106	158	140	140	168	
pH	5.7	6.2	6.1	7.2	6.3	6.6	6.5	MT4
Ec	800	381	410	320	332	109	251	
pH	5.0	5.7	6.3	6.7	6.6	7.2	7.0	MT7
Ec	350	235	190	150	141	300	216	
pH	6.7	6.5	5.9	6.4	7.2	6.8	6.4	MT15
Ec	250	186	120	250	279	200	145	

Ec : Electrical conductivity-m.mho/cm; 1982 monsoon and 1981 postmonsoon samples not analysed.

Table 1.3 : Chemical analyses of water samples from the wells (meq/l)

Ions	*Premonsoon*			*Monsoon*		*Postmonsoon*		*Well No*
	1980	*1981*	*1982*	*1980*	*1981*	*1980*	*1982*	
Ca	9.88	5.64	7.98	0.35	0.90	1.91	0.60	
Mg	13.08	8.40	11.35	0.24	0.98	1.64	0.81	
Na+K	104.40	36.97	43.49	3.91	3.35	8.87	5.80	MT1
HCO_3	17.93	9.98	5.75	2.99	1.88	3.15	1.91	
Cl	94.95	31.39	51.62	0.28	2.34	6.40	4.40	
SO_4	14.68	9.68	5.45	1.21	1.00	2.14	0.91	
Ca	1.69	1.09	1.45	1.19	1.19	1.25	1.20	
Mg	0.06	0.15	0.16	0.09	0.08	0.68	0.90	
Na+K	4.35	3.35	1.33	1.90	0.63	0.51	0.43	MT2
HCO_3	1.42	1.41	1.08	1.00	0.80	1.41	1.30	
Cl	3.27	3.27	1.59	2.00	1.00	0.71	0.80	
SO_4	1.40	0.42	0.27	0.18	0.15	0.32	0.43	
Ca	0.35	0.30	0.90	0.40	0.65	0.44	1.69	
Mg	0.05	0.06	0.008	0.04	0.14	0.04	0.10	
Na+K	0.30	0.30	0.25	0.24	0.43	0.38	0.09	MT3
HCO_3	0.36	0.36	0.87	0.38	0.78	0.50	1.58	
Cl	0.28	0.22	0.20	0.22	0.45	0.31	0.14	
SO_4	0.05	0.08	0.08	0.05	0.08	0.45	0.16	
Ca	2.39	0.80	0.90	0.95	0.94	0.55	0.41	
Mg	0.15	0.33	0.43	0.24	0.22	0.16	0.32	
Na+K	10.87	3.69	4.00	1.74	1.52	0.87	1.72	MT4
HCO_3	2.13	0.85	0.49	0.61	0.63	0.65	0.62	
Cl	9.91	3.76	3.46	1.86	1.72	0.65	1.51	
SO_4	1.37	0.21	1.28	0.47	0.33	0.28	0.30	
Ca	0.50	0.40	0.50	0.50	0.35	0.35	–	
Mg	0.24	0.16	0.24	0.16	0.16	0.57	–	
Na+K	1.37	1.33	1.13	1.25	0.43	1.68	–	MT7
HCO_3	0.52	0.33	0.28	0.33	0.13	0.95	–	
Cl	1.38	1.27	1.00	1.41	0.62	1.44	–	
SO_4	0.21	0.29	0.39	0.16	0.21	0.21	–	
Ca	1.05	0.65	0.65	0.63	0.63	1.65	–	
Mg	0.33	0.33	0.24	0.41	0.57	0.41	–	
Na+K	0.93	0.30	0.40	1.45	1.54	0.40	–	MT15
HCO_3	1.44	0.92	0.82	0.67	0.65	2.00	–	
Cl	0.84	0.28	0.39	1.72	1.85	0.34	–	
SO_4	0.05	0.08	0.10	0.10	0.28	0.11	–	

1982 monsoon and 1981 postmonsoon samples of all the wells and 1982 postmonsoon samples of wells MT7 and MT15 not analysed.

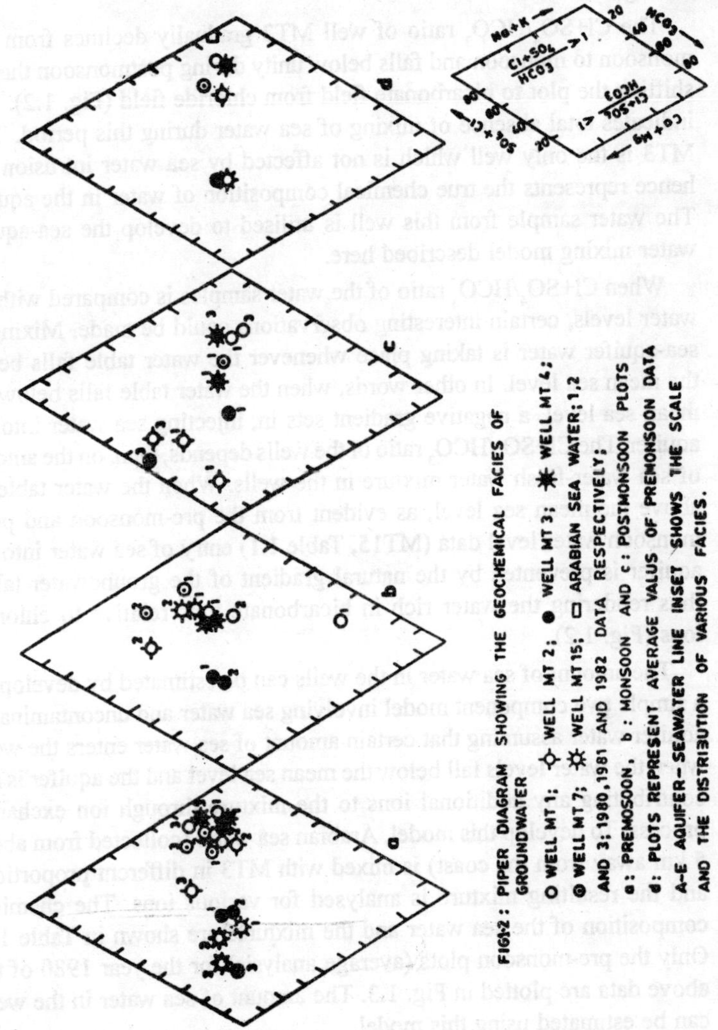

FIG.1.2 : PIPER DIAGRAM SHOWING THE GEOCHEMICAL FACIES OF GROUNDWATER
○ WELL MT 1; ⬦ WELL MT 2; ● WELL MT 3; ✱ WELL MT 4;
◎ WELL MT 7; ☼ WELL MT 15; □ ARABIAN SEA WATER; 1,2 AND 3; 1980,1981 AND 1982 DATA RESPECTIVELY;
a : PREMOSOON, b : MONSOON AND c : POSTMONSOON PLOTS
d : PLOTS REPRESENT AVERAGE VALUS OF PREMONSOON DATA
A–E AQUIFER–SEAWATER LINE INSET SHOWS THE SCALE AND THE DISTRIBUTION OF VARIOUS FACIES.

high tide and injects some amount of this mixed water into the aquifer. Due to this complex mixing phenomena, interpretation of data of this well becomes difficult. Mixing of sea water in well MT15 is noticed only during monsoon.

The $Cl+SO_4/HCO_3$ ratio of well MT2 gradually declines from pre-monsoon to monsoon and falls below unity during postmonsoon thereby shifting the plot to bicarbonate field from chloride field (Fig. 1.2). This indicates total absence of mixing of sea water during this period. Thus MT3 is the only well which is not affected by sea water intrusion and hence represents the true chemical composition of water in the aquifer. The water sample from this well is utilised to develop the sea-aquifer water mixing model described here.

When $Cl+SO_4/HCO_3$ ratio of the water samples is compared with the water levels, certain interesting observations could be made. Mixing of sea-aquifer water is taking place whenever the water table falls below the mean sea level. In other words, when the water table falls below the mean sea level, a negative gradient sets in, injecting sea water into the aquifer. The $Cl+SO_4/HCO_3$ ratio of the wells depends, then, on the amount of sea water-fresh water mixture in the wells. When the water table lie above the mean sea level, as evident from the pre-monsoon and post-monsoon water level data (MT15, Table 1.1) entry of sea water into the aquifer is prevented by the natural gradient of the groundwater table, thus rendering the water rich in bicarbonate ions relative to chloride ions (Fig. 1.2).

The amount of sea water in the wells can be estimated by developing a simple two component model involving sea water and uncontaminated aquifer water assuming that certain amount of sea water enters the wells when the water levels fall below the mean sea level and the aquifer is not contributing any additional ions to the mixture through ion exchange process. To develop this model, Arabian sea water (collected from about 5 km away from the coast) is mixed with MT3 in different proportions and the resulting mixture is analysed for various ions. The chemical composition of the sea water and the mixtures are shown in Table 1.4. Only the pre-monsoon plots (average analysis) for the year 1980 of the above data are plotted in Fig. 1.3. The amount of sea water in the wells can be estimated using this model.

The position of the plots with respect to the sea-aquifer water line suggests that the influence of clay minerals in modifying the chemical composition of the sea-aquifer water mixture is negligible. Considering the porosity and permeability of the aquifer and the density of the wells in the area studied, the water in wells MT7 and MT15, whose distance from the coast is equal to that between the well MT4 and the coast, should also show similar amount of sea water, because in coastal aquifers Ghyben-Herzberg sea water intrusion mechanism is generally invoked. If such mechanism is invoked, then, when the water levels fall below the mean sea level, the wells should contain only sea water and the plots MT4, MT7 and MT15 should shift towards the sea water in Fig. 1.3. On the otherhand, assuming a simple mixing model, the sea water intrusion mechanism, here, can be explained wherein certain amount of sea water is entering the aquifer whenever the gradient is reversed. The rate of mixing depends on the amount of salt water entering the wells which is controlled by the amount of withdrawal from the wells. Though negative gradient near well MT2 is much larger compared to the other wells, the rate of mixing is minimised because of the distance of the well from the coast. Considering the number of open wells in use, the reverse gradient is caused due to excess withdrawal of water from the wells. Such a mechanism of sea water intrusion is not uncommon in coastal aquifers (Desai *et al.*, 1979).

Table 1.4. Partial chemical analyses of sea-aquifer water mixture

	MT3	*0.5%*	*1%*	*2%*	*4%*	*6%*	*8%*	*10%*	*30%*	*100%*
EC	114	425	640	1100	1900	2850	3150	4550	12000	34000
TDS	80	297	448	770	1330	1995	2205	3105	8400	23000
Na	0.08	3.6	6.0	12.2	18.7	29.5	41.7	65.2	132.6	1130
K	–	0.2	0.3	2.9	4.1	8.2	8.7	11.5	33.2	127.8
Cl	0.51	3.9	6.9	11.5	20.3	38.0	46.5	67.7	146.6	1100

Percentage indicate amount of sea water (SW) in 100ml of SW + aquifer water mixture
MT3 : 1980 premonsoon sample from well MT3 (Aquifer water taken for mixing model)
Ec m.mho/cm: concentration in meq/l. TDS in ppm

Conclusions

1. The water table in certain wells along the Malappuram coast falls below the mean sea level (seasonal) and in certain wells it lies permanently below the sea level.
2. Groundwater occurring in the aquifer belong to bicarbonate and chloride types. Those belonging to chloride type show evidence of sea water mixing.

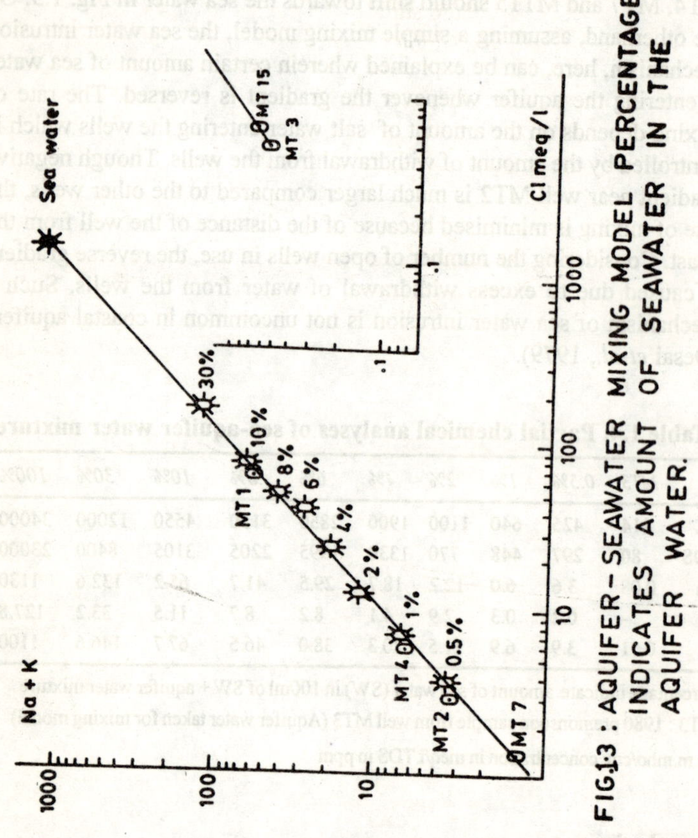

FIG.1.3 : AQUIFER – SEAWATER MIXING MODEL PERCENTAGE INDICATES AMOUNT OF SEAWATER IN THE AQUIFER WATER.

3. Correlation between the water lêvels and water chemistry shows that sea water mixing takes place whenever negative gradient sets in due to fall in water table below the mean sea level because of excess withdrawal of water from wells.
4. Sea water intrusion mechanims involved here is a simple mixing phenomena where certain amount of sea water mixes with aquifer water due to reverse in water table gradient. The amount of sea water in the wells can be estimated using a two component mixing model. Ghyben-Herzberg intrusion mechanism need not be invoked in such areas.

Acknowledgements

The author wish to thank the Director, and the Head, Department of Earth Sciences, IIT Bombay for providing facilities to process the data. The Executive Director and the Head, Groundwater Division, Centre for Water Resources Development and Management, Calicut provided facilities to carry out the work. Their help is acknowledged.

References

APHA, (1975). Standard method for examination of water and waste water. American Public Health Association, American Water Works Association and Water Pollution Control Federation.

Chandrasekharam, D., Ramesh R. and Balasubramanian, J. (1989). Geochemistry, Oxygen and Hydrogen isotope ratios of thermal springs, Western continental margin of India Field and experimental results Proceed. 6th the Water rock Interaction, Malvern, U.K. Ed. D.L. Miles, Rotterdam. U.K., pp. 149-54.

Desai, B.L. Gupta, S.K. Shah, M.V. and Sharma, S.C., (1979). Hydrological evidence of sea water intrusion along the Mangrol-Choward coast of Saurashtra, Gujrat. Hydro. Sci. Bull., 24, pp. 71-82.

Piper, A.M., (1944). A graphical procedure in the geochemical interpretation of water analyses. Am. Geophy. Union Trans., 25, pp. 914-23.

Malikarjuna, C. Vidhydharan, K.T. Pawar, S.D. Senthiappan and Francis, P.G., (1979). Geological, geochemical and geotechnical aspects of the laterites of Kerala. Proceed Seminar on laterites and lateritisation, CESS, Trivandrum.

2

Agricultural Utility of Groundwater from parts of Rewa District, Madhya Pradesh, India

YAMUNA SINGH

Introduction

Groundwater is valuable only when its quality is suitable for the purpose for which it is being explored. The suitability of natural water for a particular purpose depends upon the standards of acceptable quality for that use. Water for domestic purposes should meet the physiological needs, whereas for generation of thermal power, the quality of water should not allow wastage of heat by formation of scales and corrosion of metal parts. Water for irrigation purposes should satisfy the needs of the soil and the crops, as the liquid phase in the soil-water-plant system plays an important role in plant growth and crop production. To assess the quality of natural waters for irrigation, a number of parameters are taken into consideration. In this paper, based on physico-chemical characteristics of the groundwater, topography, soil characteristics, and hydrological conditions, an attempt has been made to evaluate usefulness of the groundwater from Rewa area for irrigation purposes.

Material and Methods

A total of 38 groundwater samples were collected from perennial wells and springs, located in sandstones, shales, and limestones. The chemical analyses of the samples were carried out in accordance with the procedure described by Rainwater and Thatcher (1960) and Vogel (1961), with an accuracy of ± 2 to ± 10%.

Results and Discussion

Physico-chemical Characteristics

Ranges of physicochemical constitutents in water samples from sandstones, shales and limestones (Fig. 2.1) are given in Table 2.1, whereas, geochemical constituents of the individual samples are illustrated in Fig. 2.2, using Stiff's (1951) graphic pattern. It is apparent from Fig. 2.2, that waters from the shales and limestones contain high concentration of cations and anions, as against low concentrations in waters from sandstones. It may, however, also be noted here that waters

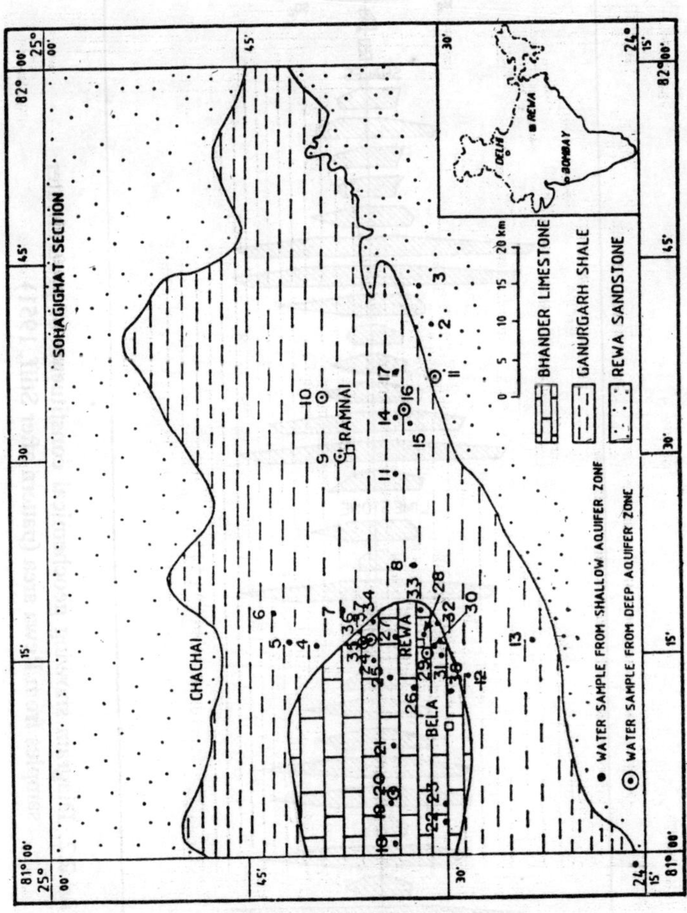

Fig. 2.1. Geological Map of the Rewa area, M.P.

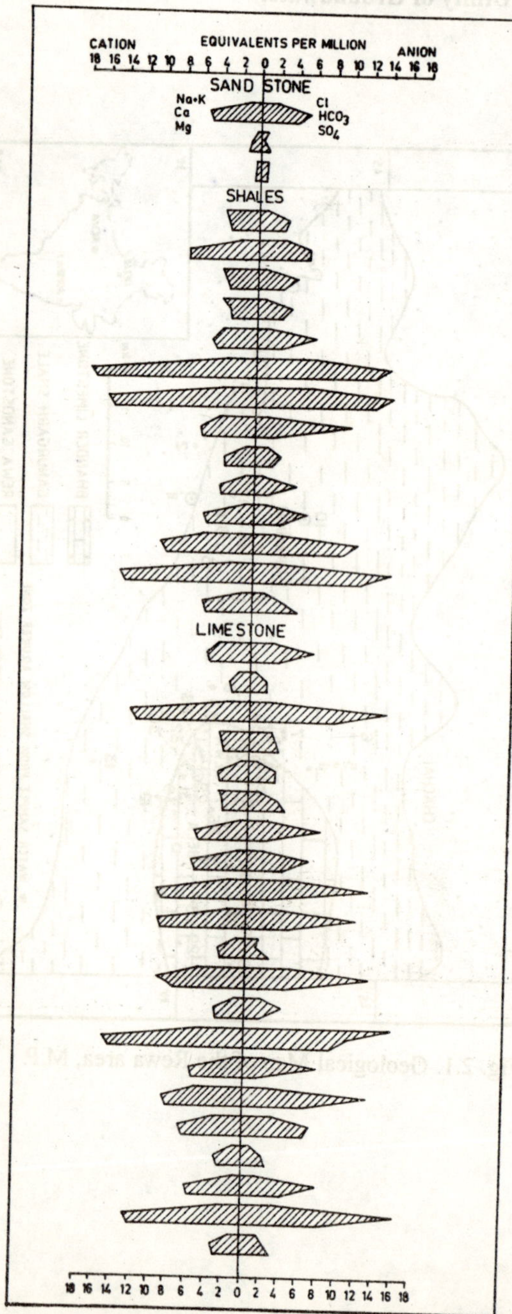

Fig. 2.2. Diagram showing geochemical constituents of groundwater samples from Rewa area (pattern after Stiff, 1951)

from deep aquifer zones are more mineralised whereas, waters from shallow aquifer zones are less mineralised (Fig. 2.3).

Table 2.1: Range of Physico-chemical constituents in groundwater samples from Rewa area

Constituents	*Rock type*					
	Sandstone		*Shale*		*Limestone*	
	A	*B*	*A*	*B*	*A*	*B*
pH	7.5	7.6	7.3	7.5	7.2	7.6
SEC	223.4	1105.3	669.6	3345.0	487.6	3356.6
Na	4.5	15.5	14.0	115.0	13.5	110.5
K	0.5	3.5	1.0	8.6	1.5	6.2
Ca	7.5	107.5	50.0	346.2	40.0	252.5
Mg	4.1	61.8	28.5	201.4	23.0	173.6
B	0.1	0.1				
F	1.5	2.0	0.5	2.5	0.5	2.0
Cl	28.6	57.2	28.6	288.7	28.6	557.7
SO_4	52.6	194.0	85.0	624.0	105.0	440.5
HCO_3	20.0	320.0	180.0	900.0	120.0	1000.0

A and B = Minimum and maximum values.

SEC = Specific electrical conductance in micro mohs/cm at 25°C.

Values of Na, K, Ca, Mg, B, F, Cl, SO_4 and HCO_3 are in parts per million.

Inspection of geochemical data of water samples, in conjunction with Palmer's (1911) scheme of water classification, indicates that in all the samples alkaline earths content exceed the bicarbonate ions and there is no residual sodium carbonate (RSC) in waters. Such a nature is further reflected by plots of the chemical data of all the samples in ABC field of Langelier and Ludwig (1942) diagram (Fig. 2.4).

Bicarbonate Hazard and Sodium Hazard

Maintenance of soil fertility depends on many factors, besides the concentration of salt in irrigation water. The salinity in soil affects crops directly more than the salinity of water. Of the many factors that control productivity, composition of irrigation water is one having cumulative effect, and therefore, various schemes have been proposed to evaluate natural waters with regards to their usefulness for irrigation purposes. Eaton (1950) proposed residual sodium carbonate (RSC) equivalents per million (epm) values to evaluate bicarbonate hazards of irrigation waters (RSC>1.25, safe; 1.25–2.5, permissible; and >2.5 unsuitable).

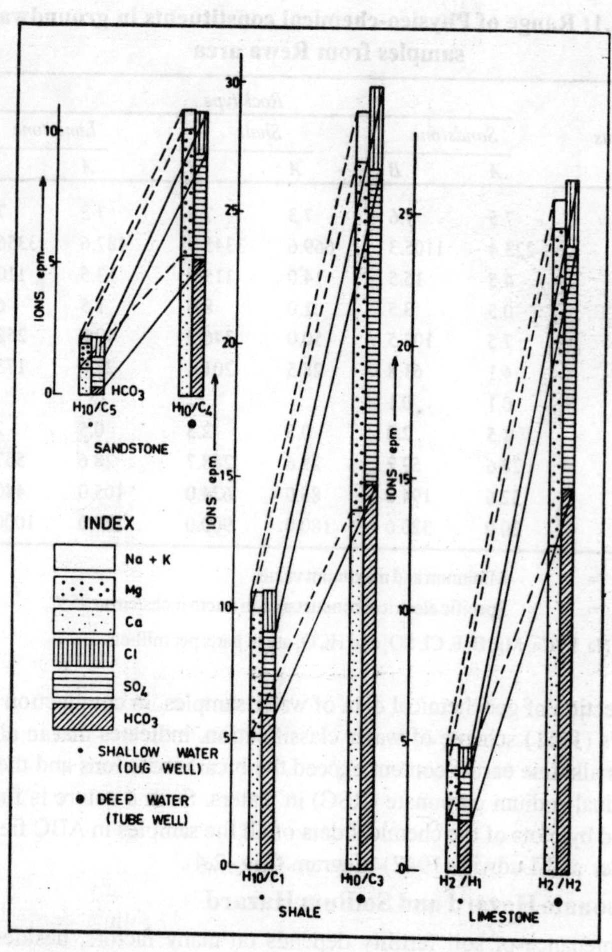

Fig. 2.3. Diagram showing mineralisation of groundwater from shallow and deep aquifer zones in Rewa area (bar after Collins, 1921).

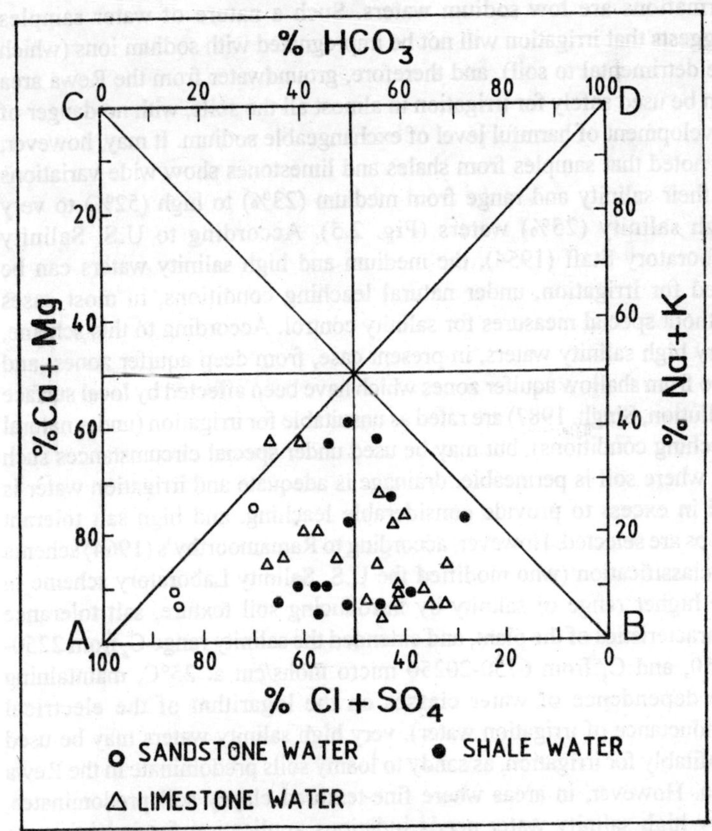

Fig. 2.4. Plots of chemical data of groundwater samples from Rewa area (field after Langelier and Ludwig, 1942).

According to this scheme, waters from Rewa area are rated as safe waters for irrigation, as RSC epm values are practically nil in all the samples.

Examination of plots of calculated values of sodium adsorption ratio (SAR) and conductivity (Fig. 2.5) indicate that samples from all the formations are low sodium waters. Such a nature of water samples suggests that irrigation will not be impregnated with sodium ions (which are detrimental to soil), and therefore, groundwater from the Rewa area can be used safely for irrigation in almost all the soils, with no danger of development of harmful level of exchangeable sodium. It may, however, be noted that samples from shales and limestones show wide variations in their salinity and range from medium (23%) to high (52%) to very high salinity (25%) waters (Fig. 2.5). According to U.S. Salinity Laboratory Staff (1954), the medium and high salinity waters can be used for irrigation, under natural leaching conditions, in most cases without special measures for salinity control. According to this scheme, very high salinity waters, in present case, from deep aquifer zones; and also from shallow aquifer zones which have been affected by local surface pollution, Singh, 1987) are rated as unsuitable for irrigation (under natural leaching conditions), but may be used under special circumstances such as, where soil is permeable, drainage is adequate and irrigation water is fed in excess to provide considerable leaching, and high salt tolerant crops are selected. However, according to Ramamoorthy's (1964) scheme of classification (who modified the U.S. Salinity Laboratory scheme in the higher range of salinity by introducing soil texture, salt tolerance characteristics of the plant, and extended the salinity range C_4 from 2250-6750, and C_5 from 6750-20250 micro mohs/cm at 25°C, maintaining the dependence of water classes on the logarithm of the electrical conductance of irrigation water), very high salinity waters may be used profitably for irrigation, as sandy to loamy soils predominate in the Rewa area. However, in areas where fine-textured clayey soil predominates, very high salinity water needs judicious application for irrigation, to avoid development of salinity hazard in soil profiles in long run.

Based on specific electrical conductance (SEC), per cent sodium (%Na), and boron concentration, Wilcox (1955), evolved another scheme of evaluating irrigation waters (Table 2.2). A comparative study of range of %Na in waters from study area (Table 2.3), against its specifications, indicates that the waters are excellent to good for irrigation. Similarly, the range of SEC, against its limits, also suggests that most of the waters from shales and limestones are good to permissible, and a few being doubtful to unsuitable for irrigation, though the waters are free from boron hazard.

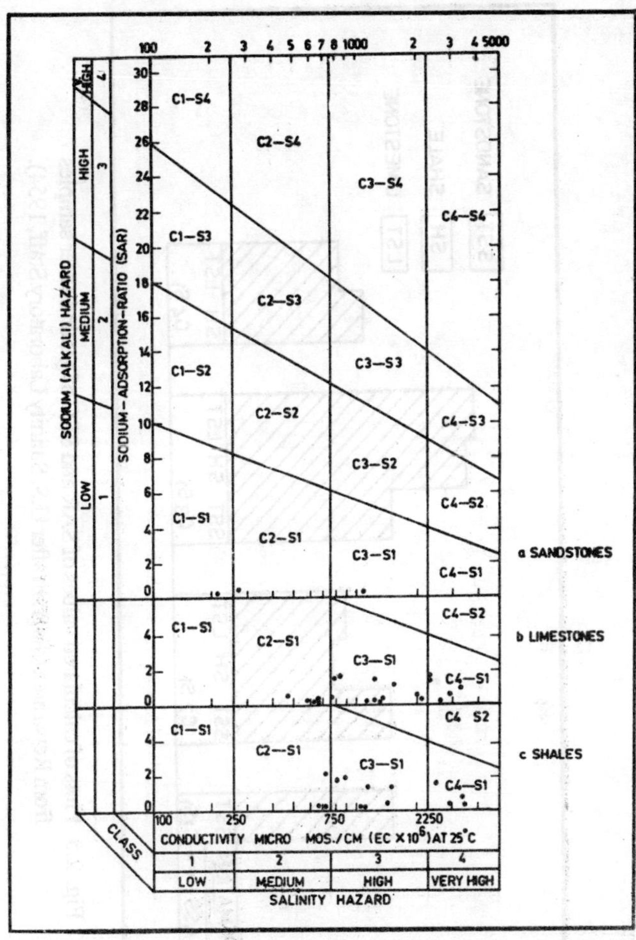

Fig. 2.5. Plots of calculated values of SAR and SEC of groundwater samples from Rewa area (diagram after U.S. Salinity Laboratory Staff, 1954).

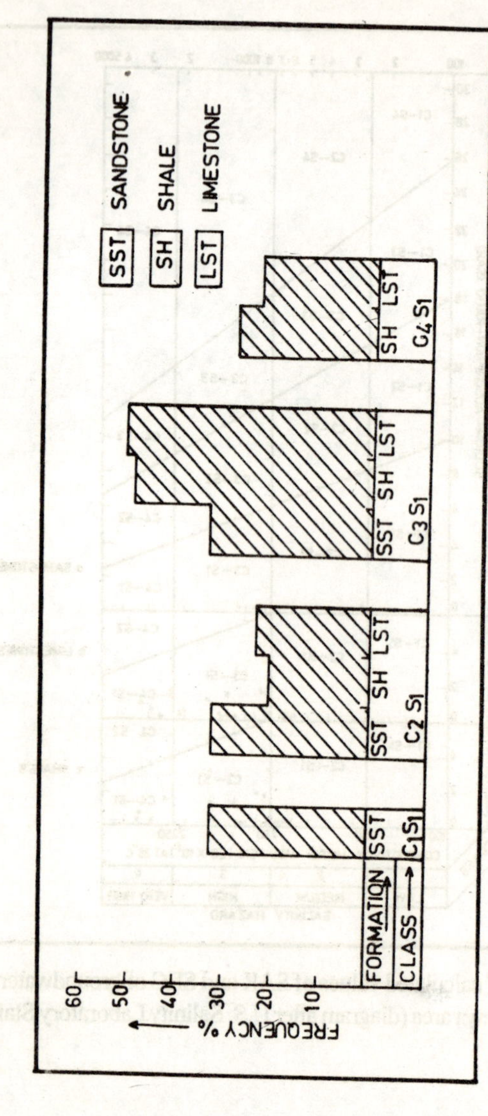

Fig. 2.5. Plots of calculated values of SAR and SEC of groundwater samples from Rewa area (diagram after U.S. Salinity Laboratory Staff, 1954).

Table 2.2 : Rating of irrigational water based on %Na, SEC and boron concentration (after Wilcox 1955)

			Boron concentration in ppm		
Water class (Rating grade)	*%Na*	*SEC*	*Sensitive crops*	*Semi-tolerant*	*Tolerant crops*
Excellent	20	250	0.33	0.67	1.0
Good	20-40	250-750	0.33-0.67	0.67-1.33	1.0-2.0
Permissible	40-60	750-2000	0.67-1.0	1.33-2.0	2.0-3.0
Doubtful	60-80	2000-3000	1.0-1.25	2.0-2.50	3.0-3.75
Unsuitable	80	3000	1.25	2.50	3.75

Magnesium Hazard

Since concentration of magnesium ions plays an important role in crop production, Paliwal (1972) used MgX100/Ca + Mg ration as an index of magnesium hazard of irrigation water. According to him, when this ratio exceeds 50, the magnesium hazard is likely to be developed in the soil profiles. As values of this ratio is below 50 in all the samples (see Table 2.3), it may be concluded that waters from study area are free from magnesium hazard and can be used for irrigation.

Chloride Ion Effect

Similar to magensium ions, chloride ions concentrations also deserve attention as according to Ayers and Branson (1975), when epm value of chloride in irrigation water is more than 10, water is likely to pose severe problem and affect crop production adversely. Since epm values of chloride ions, in present case, are less than 10, groundwater is rated as suitable for irrigation.

Topography, Soils and Hydrological Conditions

An undulating topography of the Rewa area helps quick removal of water by overland flow. Similarly, well developed natural drainage in the area also drains water, received from rainfall (114.34 cm, average for 56 years). These features do not allow prolonged stagnation of water (and its longer contact with the underlying soils) and therefore, there is no chance for significant accumulation of salts in the soils profiles. It may be further noted here that at no time water table is close to surface in Rewa area, as evidenced by pre- and post-monsoon water level measurements (Singh, 1985). Such a hydrological condition in the area provides free circulation of water in the aeration zone and helps removal of salts from the soil profiles, if accumulated. It means that enough natural leaching occurs in the Rewa area and therefore, even if saline water is

Table 2.3 : Calculated values of SAR, %Na, Mg hazard and RSC* in groundwater samples from Rewa area

Sample No.	*SAR*	*%Na*	*Mg hazard*
1	2	3	4
1.	0.30	6.79	48.66
2.	0.20	9.63	49.24
3.	0.39	25.26	47.89
4.	1.88	36.69	47.66
5.	0.58	10.08	49.68
6.	1.99	37.76	48.53
7.	2.16	41.40	48.35
8.	1.53	27.38	48.34
9.	0.42	5.24	48.91
10.	0.84	10.05	48.63
11.	1.41	23.46	48.67
12.	0.37	10.82	48.45
13.	0.33	9.43	48.68
14.	0.28	6.75	48.64
15.	1.64	21.68	48.55
16.	0.47	6.78	48.87
17.	0.28	6.87	48.38
18.	1.56	27.49	48.57
19.	0.66	20.45	48.59
20.	0.30	4.39	48.76
21.	1.90	37.02	48.72
22.	0.37	10.68	48.63
23.	1.86	37.05	48.93
24.	0.26	6.38	48.75
25.	0.28	6.61	48.57
26.	0.61	9.45	48.73
27.	0.57	8.98	48.59
28.	0.54	14.77	48.19
29.	1.57	21.73	45.94
30.	0.53	14.09	48.41
31.	1.03	13.77	48.82
32.	0.25	5.72	48.80
33.	1.66	22.88	48.60
34.	1.23	19.57	48.59
35.	0.54	15.12	47.82

1	2	3	4
36.	0.28	6.41	48.67
37.	0.68	9.17	48.72
38.	0.52	14.22	48.36

1 to 3 : Samples from sandstones

4 to 17 : Samples from shales

18 to 38 : Samples from limestones.

* Values of RSC is zero in all the samples except in sample No. 3 where it is 0.27.

applied for irrigation, soil profiles are not likely to develop salinity hazard. However, in areas where fine-textured clayey soil predominates and inhibits downward movement of the water due to its high water retention capacity, application of very high salinity water for irrigation should be violated to avoid development of salinity in soil profiles in long run.

Conclusion

(1) In the Rewa area of Madhya Pradesh, waters from shales and limestones contain high concentration of cations and anions. It is also observed that groundwater is more mineralised in deep aquifer zones, as against lesss mineralised water in shallow aquifer zones, irrespective of lithology.

(2) Groundwater from shales and limestones are mostly medium to high salinity waters, and can be used profitably for irrigation, in all types of soils in the Rewa area, under natural leaching conditions, topography, soil characteristics and hydrological conditions.

(3) Very high salinity water (deep aquifer zone water) needs judicious applications for irrigation purposes in areas where fine-textured clayey soil predominates, to avoid development of salinity hazard in soil profiles in long run.

(4) Fine-textured clayey soil should be irrigated by making use of groundwater from shallow aquifer zones.

Acknowledgement

The author thanks Shri Y.V.S. Rao for drafting figures and S. Rajaiah for typing the manuscript.

References

Ayers, R.S. and Branson, R.L., (1975). Guidelines for interpretation of water quality for agriculture. Univ. of California Extension Mimeographed, 13 pp.

Collins, W.D., (1923). Graphical representation of water analysis. Ind. Eng. Chem., V. 15.

Eaton, F.M., (1950). Significance of carbonates in irrigation water. Soil Science, V. 69, pp. 123-33.

Langelier, W.F. and Ludwig, H.F., (1942). Graphical methods for indicating the mineral character of natural waters. Jour. Amer. Water Works Assoc., V. 34, no. 3, pp. 335-52.

Paliwal, K.V., (1972). Irrigation with saline water. I.A.R.I. Monograph No. 2 (New Series), New Delhi, 198 pp.

Palmer, C., (1911). The geochemical interpretation of water analysis. U.S.G.S. Bull. 479, pp. 1-19.

Rainwater, F.H. and Thatcher, L.L., (1960). Methods for collection and analysis of water samples U.S.G.S. Water Supply Paper–1454, 301 pp.

Ramamoorthy, B., (1964). Some aspects of management of soil water plant climatic complex in the arid zone. Proc. General Symp. on Problems on Indian Arid Zones. CAZRI, Jodhpur, pp. 276-85.

Singh, Y., (1985). Hydrogeology of the karstic area around Rewa, Madhya Pradesh, India. Proc. at International Seminar on Karst Water Resources, Ankara, Antalya, IAHS Publ. No. 161, pp. 407-16.

Singh, Y., (1987). Groundwater pollution in karst aquifers near Rewa town, Madhya Pradesh, *In* Sahu, K.C. (Editor), Proc. National Symp. Role of Earth Sciences in Environment, Bombay (1987), pp. 293-97.

Stiff, H.A. Jr., (1951). The interpretation of chemical analysis by means of patterns. J. Petrol. Tech. pt. 15.

U.S. Salinity Laboratory Staff, (1954). Diagnosis and improvement of saline and alkali soils. U.S.D.A. Handbook no. 60, 160 pp.

Vogel, A.I., (1961). A Text Book of Quantitative Inorganic Analysis, Longmans, 3rd edition, London.

Wilcox, L.V., (1955). Classification and use of irrigation water U.S.D.A. Circ., 969, Washington, D.C., 19 pp.

3

Irrigational Quality of Groundwater from Pliocene and lower Miocene Aquifers of Cauvery Delta, Tamil Nadu

S. SATHYAMOORTHY AND N. DHAKSHINAMOORTHY

Introduction

The Cauvery delta, encompassing almost 8,000 sq. km. of land is the main area of rice production in Tamil Nadu. The delta is irrigated mainly by water received from the controlled release at Mettrur in Cauvery replenished by flood water derived from the southwest monsoon. In recent years the water supply from the Cauvery has become erratic. The rainfall is irregular and undependable and the development of groundwater is become necessary. The objective of this work is to access the irrigational quality of groundwater from two selected sub-surface aquifers of the Cauvery Delta.

Cauvery Delta

The Cauvery delta is located in the southern part of Tamil Nadu (Fig. 3.1) and confluences the sea between 10°15', and 11°20' N latitudes. It is bounded by the Coleroon river in the north, the Grand Anicut canal in the west and to the south and east by Palk Strait and Bay of Bengal respectively.

The delta with a slightly undulating terrain slopes gradually from an altitude of about 50m. in its western part to sea level at its southern and eastern limits. The area has, in general, humid tropical climate with an average annual rainfall 1188 mm. The delta area constitutes about 80 per cent of the Thanjavur District. Most of the people are engaged in farming and the production of rice is the mainstay of the District.

Sub-Basins

Based on the lithologic and hydrogeologic conditions the Cauvery Delta has been divided into three sub-basins, namely : (1) Cauvery sub-basin, (2) Vennar sub-basin and (3) New Delta area (Fig. 3.2).

The Cauvery sub-basin has a flat topography and is underlain mainly by Quaternary deposits of fluvial origin which grade into marine facies

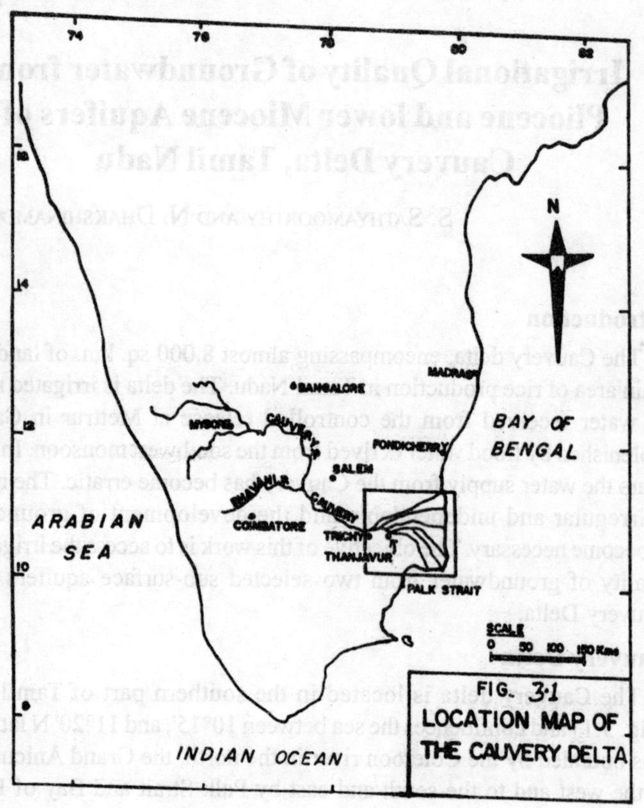

FIG. 3·1 LOCATION MAP OF THE CAUVERY DELTA

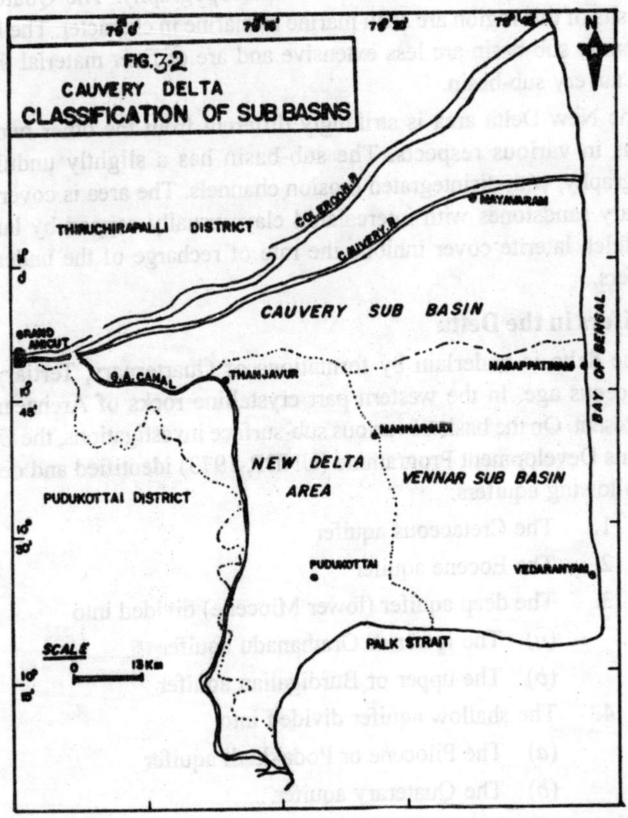
FIG. 3.2
CAUVERY DELTA
CLASSIFICATION OF SUB BASINS
N
THIRUCHIRAPALLI DISTRICT
COLEROON R
CAUVERY R
MAYAVARAM
CAUVERY SUB BASIN
GRAND ANICUT
G.A. CANAL
THANJAVUR
NAGAPPATINAM
BAY OF BENGAL
MANNARGUDI
NEW DELTA AREA
VENNAR SUB BASIN
PUDUKOTTAI DISTRICT
PUDUKOTTAI
VEDARANYAM
PALK STRAIT
SCALE
0
13 Km

east of Mayavaram. Extensive natural levee complexes provide an excellent shallow aquifer system in this sub-basin.

The Vennar sub-basin also has a flat topography. The Quaternary deposits of this region are semi marine to marine in character. The levees in Vennar sub-basin are less extensive and are of finer material than in the Cauvery sub-basin.

The New Delta area is strikingly different from the other two sub-basins in various respects. The sub-basin has a slightly undulating topography, with disintegrated erosion channels. The area is covered by Tertiary sandstones with intercalated clays usually capped by laterite. The thick laterite cover inhibits the rate of recharge of the underlying aquifers.

Aquifers in the Delta

The delta is underlain by formations of Quarternary, Tertiary and Cretaceous age. In the western part crystalline rocks of Archaean age are present. On the basis of various sub-surface investigations, the United Nations Development Programme (UNDP, 1973) identified and defined the following aquifers:

1. The Cretaceous aquifer
2. The Eocene aquifer
3. The deep aquifer (lower Miocene) divided into
   (*a*) The lower or Orathanadu aquifer
   (*b*) The upper or Burdigalian aquifer
4. The shallow aquifer divided into
   (*a*) The Pilocene or Podakkudi aquifer
   (*b*) The Quaterary aquifer.

In the present study the irrigational quality of water from one shallow aquifer (Pilocene aquifer) and one deep aquifer (Lower Miocene-Orathanadu aquifer) has been studied. Pilocene and Miocene outcrops are exposed on the western side of the delta and have a gentle dip towards east. The Pilocene aquifer is composed of sandy deposits with major clay intercalations towards the east. Its thickness increases from 50m. in the New Delta area to about 75m. along the coast, in depth ranging from 45 to 115m. The Lower Miocene aquifer is made of sandy gravelly deposits typical of deltaic facies. Its thickness increases from 33m. near Thanajavur in the west to about 120m. along the coast. Where it occurs at depths ranging from 420 to 540 m bgl.

Materials and Methods

Chemical analysis data (partial) pertaining to 30 groundwater samples from the Pliocene aquifer and 21 samples from the Lower Miocene aquifer were made available by the Groundwater Division of the Public Works Department, Government of Tamil Nadu, Madras. The relevant data are presented in Tables 3.1 and 3.2.

The quality of an irrigation water is judged not only by the total but also by the kind of salts it contains and the individual ions involved. Most of the classifications of irrigation water. (USSL, 1954; Wilcox, 1955) supplies consider the Total Dissolved Solids (TDS) (salinity hazard) and concentrations of sodium (alkali hazard). The Sodium Adsorption Ratio (SAR) recommended by Richards (1954) has been computed for all the samples for the assessment of alkali hazard in this study.

A simple diagram for evaluation of irrigation waters on the basis of Electrical Conductivity and Sodium Adsorption Ratio has been suggested by the U.S. Salinity Laboratory (1954). The plots of groundwater from Pliocene and Lower Miocene aquifers of the Cauvery Delta in the USSL diagram are shown in Fig. 3.3 and 3.4.

Results and Discussion

In the heart of the Cauvery Delta, the soils in general are considered to be medium textured and moderately leached with adequate drainage. But the eastern part of Vennar sub-basin has a fine soil type and poor drainage. The paddy crops which constitute the mainstay of Cauvery Delta are semi-salt tolerant. Hence a salinity hazard up to C3 and alkali hazard up to S2 are usually considered safe for irrigation in this region. In other words, the classes C1S1, C1S2, C2S1, C2S2, C3S1 and C3S2 of the USSL classifications are suitable for paddy cultivation in Cauvery Delta.

In the Figs. 3.3 and 3.4 all the wells in the New Delta are tapping the Pliocene and Lower Miocene aquifers fall within the above six permissible classes suited for paddy cultivation. In the Vennar sub-basin most of the wells belonging to both the aquifers are decidely bad. But in the Cauvery sub-basin the plots of poor quality have been found to be from the wells situated towards the eastern part of the basin.

The UNDP (*op. cit.*) has proposed the quality criteria for irrigation waters for growing paddy in typical deltaic soils of Thanjavur District. (see p. 37).

Groundwater with Ec>4,000 mmhos and SAR>18.5 is considered unsuitable for paddy cultivation. Following these criteria thematic maps have been presented (Figs. 3.5 and 3.6) to bring out the water quality ratings of the Pliocene and Lower Miocene aquifers of the Cauvery delta.

Table 3.1 : Chemical Data for Groundwater from Pliocene Aquifer, Cauvery Delta

Sl.	Location Name	Ec	pH	Ca	Mg	Na + K	BCO_3	CO_3	Cl	NO_3	SO_4	SAR	USSL	Sub Basin
						ppm	connc							
1	Nirattannallur	896	9	52	20	123	323	0	99	0	53	4	C_3S_1	CSB
2	Papagudi	1,232	8	44	23	213	470	0	128	0	62	6	C_3S_2	CSB
3	Marudandanallur	588	8	22	12	107	336	0	36	0	2	4	C_2S_1	CSB
4	Peralam	1,064	8	10	11	254	555	0	64	0	37	13	C_3S_3	CSB
5	Kadambangudi	784	8	16	11	181	451	0	36	0	32	8	C_2S_2	CSB
6	Maharajapuram	1,110	9	10	13	278	476	16	89	0	53	13	C_3S_3	CSB
7	Kovilkandankudi	1,456	8	20	195	1,116	354	0	39	0	24	20	C_3S_4	CSB
8	Kiranur	784	9	10	4	186	372	24	57	0	13	13	C_2S_3	CSB
9	Kamakshipuram	1,232	9	24	6	293	634	30	30	0	61	14	C_3S_3	CSB
10	Pattamangalam	616	8	44	23	61	348	0	32	0	0	18	C_2S_1	CSB
11	Pattavartti	3,360	9	46	67	626	384	24	569	0	835	14	C_4S_4	CSB
12	Siva nagaram	528	8	40	20	50	275	0	39	0	10	2	C_2S_1	CSB
13	Valangiman	834	8	30	17	140	366	24	64	0	19	5	C_3S_1	CSB
14	Podakkudi	1,456	8	88	33	174	409	0	234	0	57	4	C_3S_1	CSB
15	Vilakudi	5,040	8	176	127	788	525	0	1,463	0	202	11	C_4S_2	VSB
16	Kilvelur	5,040	9	28	41	1,230	781	54	1,063	0	514	34	C_4S_4	VSB
17	Thirunayappeyar	4,480	8	54	89	922	561	0	1,148	0	372	18	C_4S_4	VSB
18	Kachanam	6,720	8	84	76	1,273	580	0	1,789	0	312	24	C_4S_4	VSB
19	Vedapadimangalam	7,840	8	304	200	992	470	0	2,016	0	432	11	C_4S_4	VSB
20	Kovilokalappal	11,100	8	260	256	2,532	592	0	4,083	0	854	26	C_4S_4	VSB
21	Perugavalandan	4,450	7	148	78	701	476	0	1,008	0	372	11	C_4S_3	VSB

22	Adirampatnam	1,736	9	16	2	400	476	30	234	0	92	24	C_3S_4	NDA
23	Saliyamangalam	224	9	26	9	20	85	12	25	0	17	1	C_1S_1	NDA
24	Madukku	1,232	8	90	39	121	336	0	202	0	71	3	C_3S_1	NDA
25	Thamarankottai	500	9	42	13	63	226	0	60	0	22	2	C_2S_1	NDA
26	Vadaseri	616	8	48	20	59	201	0	86	0	31	2	C_2S_1	NDA
27	Vettikkadu	896	8	64	26	122	458	0	78	0	38	3	C_3S_1	NDA
28	Sadiyarkovil	250	8	26	7	20	129	0	17	0	13	1	C_1S_1	NDA
29	Kovilvenni	778	8	70	16	72	297	0	75	0	38	2	C_3S_1	NDA
30	Okkanavukilaiyur	779	7	68	244	56	165	0	156	0	20	1	C_1S_1	NDA

Source : Public Works Department; Tamilnadu

All the concentrations are in ppm, otherwise mentioned

CSB – Cauvery Sub-basin; VSB – Vennar Sub-basin,

NDA – New Delta area

Table 3.2 : Chemical Data for Groundwater from Lower Miocene Aquifer, Cauvery Delta

Sl.	Location Name	Ec	pH	Ca	Mg	Na + K ppm	BCO_3 connc	CO_3	Cl	NO_3	SO_4	SAR	USSL	SUB Basin
1	Sannanllur	1,680	9	8	4	374	354	30	337	0	10	27	C_3S_3	CSB
2	Kiranur	1,008	8	20	9	195	287	0	153	0	49	89	C_3S_2	CSB
3	Pattamangalam	930	8	24	9	183	256	0	178	0	42	8	C_3S_2	CSB
4	Pattavartti	1,344	8	62	27	205	397	0	220	0	66	6	C_3S_2	CSB
5	Maharajapur	1,232	8	64	39	118	336	0	167	0	36	22	C_3S_1	CSB
6	Sivanaragaram	1,344	8	80	24	178	354	0	241	0	57	4	C_3S_1	CSB
7	Podakkudi	1,450	8	88	33	181	409	0	234	0	72	4	C_3S_1	VSB
8	Vilakkudi	17,000	8	681	438	3,118	268	0	6,099	0	9	23	C_4S_4	VSB
9	Kilvelur	1,670	8	22	11	363	372	0	348	0	57	16	C_3S_3	VSB
10	Vadapadimangalam	4,480	8	194	82	617	171	0	1,225	0	178	9	C_4S_3	VSB
11	Vilamal	1,344	8	20	6	322	372	36	241	0	1	16	C_3S_3	VSB
12	Adangudi	1,456	8	76	39	202	384	18	220	0	80	5	C_3S_1	VSB
13	Vilakudi	3,220	9	24	4	673	305	36	852	0	10	33	C_4S_4	VSB
14	Orathanadu	448	8	381	9	39	153	0	46	0	14	2	C_2S_1	NDA
15	Kaduvaranpatti	212	7	22	4	17	67	0	21	0	7	1	C_4S_1	NDA
16	Vadaseri	2,016	8	102	42	272	397	0	440	0	66	6	C_3S_2	NDA
17	Kovilverni	778	8	60	33	95	329	0	99	0	25	3	C_3S_1	NDA
18	Sadiyakovil	449	8	44	15	32	195	0	46	0	8	1	C_1S_1	NDA
19	Okkanadniviliangur	890	8	60	18	106	287	0	131	0	21	3	C_3S_1	NDA
20	Nayakkarpatti West	222	7	36	2	15	179	0	156	0	3	1	C_1S_1	NDA
21	Kakkarai	528	8	60	16	29	268	0	35	0	5	1	C_2S_1	NDA

Source : Public Works Department, Tamil Nadu

All the concentrations are in ppm, otherwise mentioned

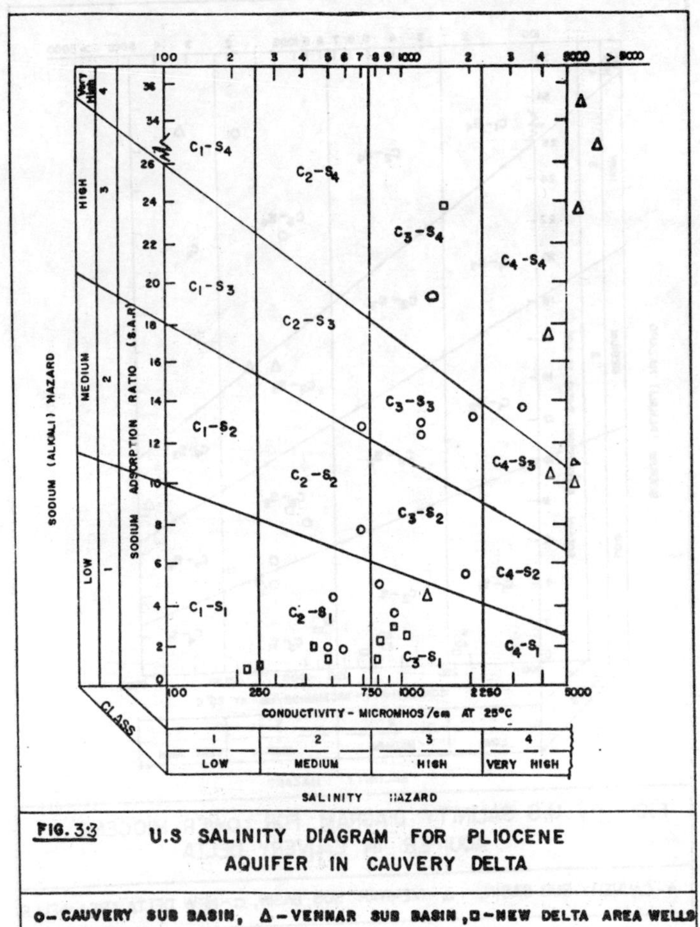

FIG. 3.3 U.S SALINITY DIAGRAM FOR PLIOCENE AQUIFER IN CAUVERY DELTA

o - CAUVERY SUB BASIN, Δ - VENNAR SUB BASIN, □ - NEW DELTA AREA WELLS

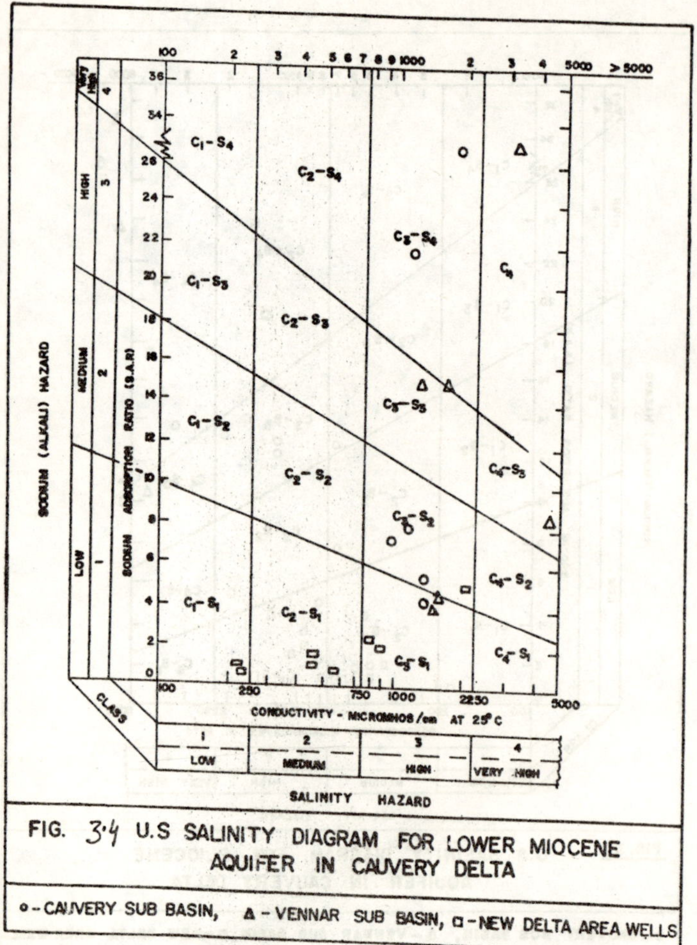

FIG. 3·4 U.S SALINITY DIAGRAM FOR LOWER MIOCENE AQUIFER IN CAUVERY DELTA

o - CAUVERY SUB BASIN, Δ - VENNAR SUB BASIN, □ - NEW DELTA AREA WELLS

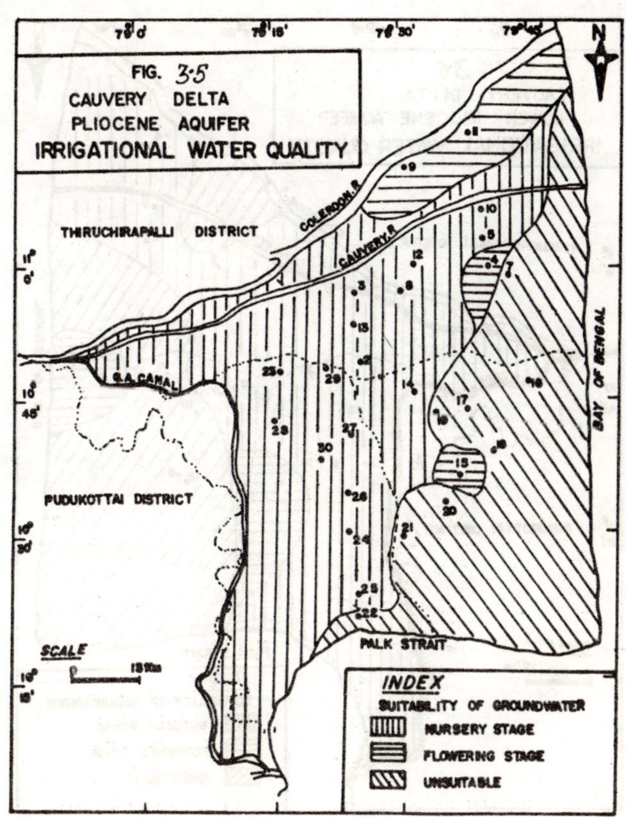

FIG. 3·5
CAUVERY DELTA
PLIOCENE AQUIFER
IRRIGATIONAL WATER QUALITY
THIRUCHIRAPALLI DISTRICT
PUDUKOTTAI DISTRICT
COLEROON R
CAUVERY R
G.A. CANAL
PALK STRAIT
BAY OF BENGAL
SCALE
INDEX
SUITABILITY OF GROUNDWATER
NURSERY STAGE
FLOWERING STAGE
UNSUITABLE

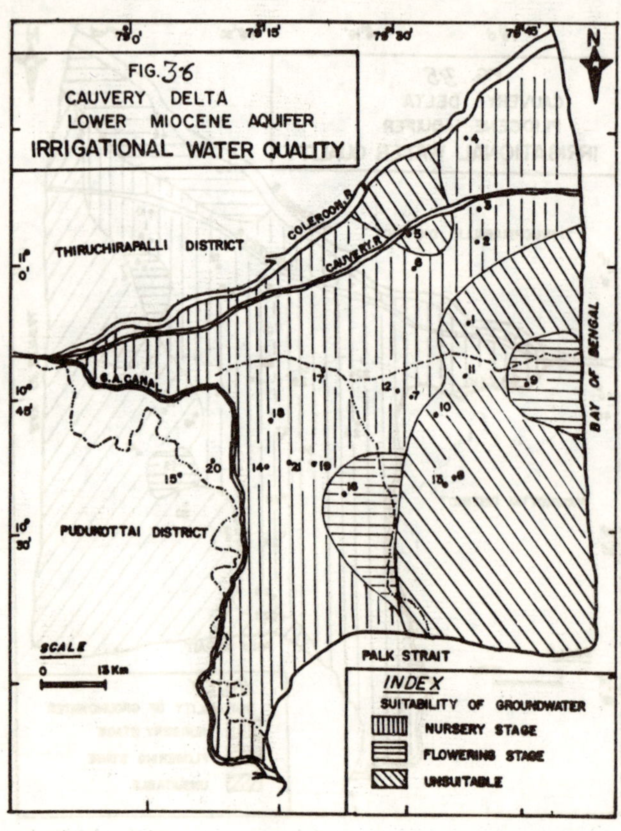
79°0'
79°15'
79°30'
79°45'
N
FIG. 3·6
CAUVERY DELTA
LOWER MIOCENE AQUIFER
IRRIGATIONAL WATER QUALITY
COLEROON. R
CAUVERY. R
THIRUCHIRAPALLI DISTRICT
11° 0'
10° 45'
10° 30'
G.A. CANAL
PUDUKOTTAI DISTRICT
BAY OF BENGAL
PALK STRAIT
SCALE
0 15 Km
INDEX
SUITABILITY OF GROUNDWATER
NURSERY STAGE
FLOWERING STAGE
UNSUITABLE

Sl. No	*Electrical Conductivity (m mhos)*	*SAR*	*Utility*
1.	up to 1,500	< 10	suitable for raising nurseries
2.	up to 3,000	< 5	–do–
3.	up to 4,000	up to 18.5	suitable for maturing rice from flowering stage

Scope for Development

The Pliocene and Lower Miocene aquifers have not been utilised in the Cauvery sub-basin since Quatenary aquifer provides groundwater at shallow depths in these areas. However, shallow wells up to a maximum depth of 35m. tapping the Pliocene aquifers are common in New Delta area. The Pliocene aquifers, have also been developed to a small extent in the western part of Vennar sub-basin around Vadapadhimangalam.

The groundwater exploitation from Lower Miocene aquifer is very meagre in the Cauvery Delta. Drilling up to a minimum depth of 100m. on the western side (Nayakkarpatti) and 180m. on the eastern side (Kakrarai) of the New Delta is necessary to tap the Lower Miocene aquifer. This can not be afforded by riots with limited land-holding and hence the State Government should come forward to develop the Miocene aquifer in the New Delta area, and provide groundwater to small farmers at a nominal cost.

Acknowledgements

The authors are indebted to their respective Institutions for the facilities provided for carrying out this work. The source materials provided by the Chief Engineer (Groundwater), PWD, Government of Tamil Nadu, are gratefully acknowledged. The authors are also thankful to Messrs. S. Subramanian, S.N. Mariappan, T. Mali and S. Kadarkarai, research staff of the Department of Geology, V.O.C. College, Tuticorin for their assistance in the preparation of the manuscript.

References

Richards, L.A., (USSL), 1954. Diagnosis and Improvement of saline and alkali soils, USDA hand book 60, 160 p.

UNDP., 1973. Groundwater investigation in Tamil Nadu (Phase II), Vol. 1 and 2.

Wilcox, L.V., 1955. Classification and use of Irrigation water, USDA Circ. 969, 19 p.

Groundwater Quality—A Case Study of Textile Industrial City, Coimbatore, India

N Someswara Rao, TNV Venkateswara Rao,
D Jagannadha Rao and B Narasimha Murthy

Introduction

Coimbatore, latitude 11°0'N and longitude 76°58'E (Fig. 4.1) is one of the major textile industrial centres in India, with approximately 150 textile mills and 100 small dyeing units. In the city area, the groundwater is very hard whereas in the surrounding villages, it is normal in quality. Seventy years ago, groundwater was the only source for domestic and other needs. With time, it became unfit for human consumption due to high hardness and salinity. The entire city, now depends for domestic and industrial needs on surface water, drawn from the Siruvani reservoir, 40 km. away from the city. No attempt has been undertaken to study the gradual deterioration of the groundwater quality with the contamination of the textile mills effluents. Hence, an attempt is made to study the impact of these effluents on groundwater quality in Coimbatore city.

Material and Methods

Sixty water samples were collected from open wells, bore wells, the source reservoir and the consumer ends of distribution system. The samples were preserved by adopting standard techniques (Franson, 1985) such that no significant changes in the composition occur before the analysis. Various chemical parameters were analysed by using standard methods (Franson, 1985; Golterman, 1978; Vogel, 1978; and Jan and Young, 1978).

Results and Discussion

The water samples and industrial effluents were analysed for various chemical parameters and the data is presented in Tables 4.1, and 4.2 respectively. From the data, the following points can be inferred.

(*a*) the groundwater in the city area is very hard.

(*b*) the groundwater in the villages, surround the city, is normal with reference to hardness.

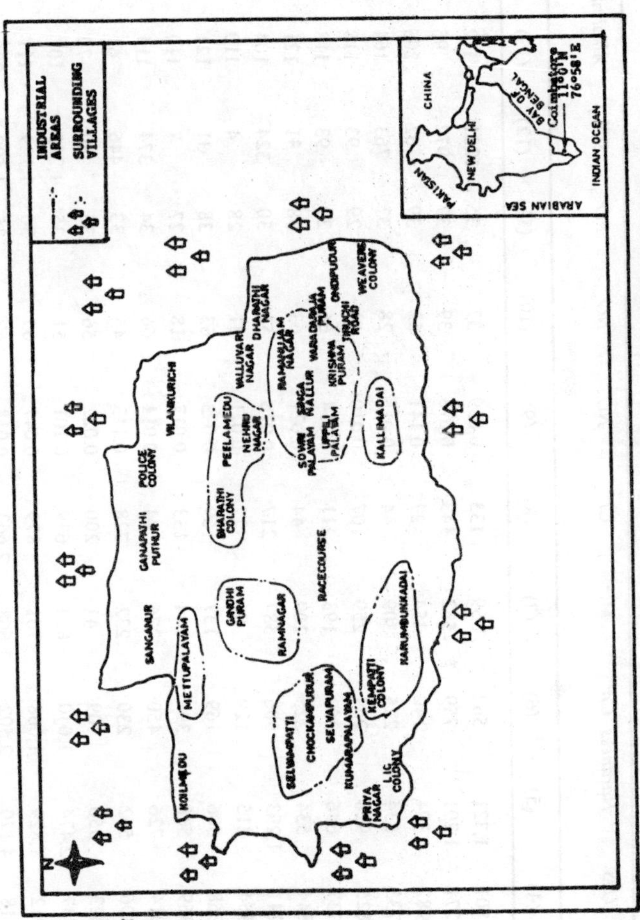

FIG.4.1. MAP SHOWING THE AREAS FROM WHERE SAMPLES HAVE BEEN COLLECTED

Table 4.1 : Analysis of Ground Water in Coimbatore City

(in mg/l)

Sample No.	pH	E.C**	TDS	Hardness	Ca^{2+}*	Mg^{2+}*	Cl^-	NO_2^-	NO_3^-	PO^{3-}_4	SO^{2-}_4	Alkalinity	F^-
(1)	(2)	(3)	(4)	(5)	(6)	(7)	(8)	(9)	(10)	(11)	(12)	(13)	(14)
1.	2.96	5.20	2,801	1,321	562	559	133	0.039	37	35	1231	112	0.65
2.	8.31	6.99	4,271	1,291	759	532	143	0.073	39	37	972	98	1.33
3.	8.26	1.74	982	394	140	254	27	0.141	52	39	62	206	0.81
4.	8.33	3.56	1,735	534	225	309	44	0.015	28	37	763	164	2.47
5.	8.17	3.27	1,825	928	269	259	107	0.010	64	29	93	118	0.15
6.	7.57	2.90	1,737	646	253	193	111	0.014	64	29	93	118	0.15
7.	8.19	1.76	844	534	225	209	44	0.021	63	38	31	121	0.82
8.	8.20	4.65	3,241	1,293	506	587	217	0.078	51	30	324	104	1.17
9.	8.43	1.05	854	215	129	86	27	0.011	37	28	4	110	0.86
10.	8.12	3.77	2,068	506	169	137	169	0.033	63	38	31	121	0.82
11.	7.80	2.93	1,369	538	394	144	133	0.017	18	27	7	110	0.43
12.	2.93	3.11	2,014	726	450	276	84	0.014	66	34	374	114	0.37
13.	7.98	4.29	2,806	562	230	232	728	0.113	45	33	446	63	0.52
14.	8.41	1.99	1,223	350	309	41	200	0.022	56	35	80	70	0.88
15.	8.04	5.95	3,747	2,474	1,040	834	639	0.101	51	35	1,255	107	0.38
16.	7.93	6.49	4,912	1,829	1,068	761	879	0.047	69	36	1,129	111	0.60
17.	7.88	12.04	10,118	4,170	2,502	1,668	2,663	0.018	67	35	1,590	220	0.46

18.	7.36	9.67	6,622	2,099	1,249	850	1,775	0.062	66	36	1,471	12	0.53
19.	7.92	5.72	3,330	1,490	618	672	817	0.683	65	36	647	142	0.50
20.	7.66	6.26	2,752	984	647	337	1,207	0.028	57	36	333	121	0.50
21.	7.48	2.81	2,125	1,211	365	266	24	59	35	32	236	0.41	
22.	7.12	4.65	3,069	843	782	761	905	0.019	43	30	81	159	0.48
23.	7.23	2.47	1,328	537	337	201	302	0.018	37	29	21	161	0.73
24.	7.24	3.72	1,993	1,063	675	388	604	0.018	55	27	69	110	0.27
25.	7.23	3.11	1,572	815	506	309	426	0.017	56	36	45	128	0.70
26.	7.47	3.08	1,483	590	362	228	444	0.022	66	38	55	93	1.16
27.	8.20	3.24	1,608	675	423	252	426	0.019	60	34	169	142	1.73
28.	8.03	2.96	1,194	562	351	211	408	0.051	57	33	67	93	1.08
29.	7.38	1.84	942	534	224	200	142	0.032	66	37	28	124	0.75
30.	7.72	1.50	1,160	590	368	222	178	0.035	29	24	11	131	0.30
31.	7.60	3.39	1,842	888	552	336	497	0.031	67	36	36	96	0.82
32.	8.64	1.67	984	506	314	192	124	0.025	17	27	57	119	0.30
33.	7.76	3.35	2,494	984	606	378	517	0.285	64	29	63	135	0.39
34.	7.26	2.23	1,144	877	217	160	249	0.024	49	29	62	174	0.63
35.	7.77	2.33	1,175	562	351	211	284	0.209	59	29	62	158	0.87
36.	7.47	3.17	1,353	675	412	263	213	0.032	60	36	149	166	4.65
37.	7.36	3.92	2,411	787	491	296	266	0.127	61	37	675	198	1.63
38.	7.52	3.33	1,770	731	437	294	373	0.061	62	37	504	206	1.13
39.	7.63	2.13	1,694	606	325	181	108	0.036	60	35	10	245	1.53
40.	8.28	1.77	1,234	534	297	237	107	0.013	48	34	40	182	0.76

(Contd.)

(1)	(2)	(3)	(4)	(5)	(6)	(7)	(8)	(9)	(10)	(11)	(12)	(13)	(14)
41.	8.23	2.65	1,440	872	542	330	391	0.070	34	30	42	127	2.44
42.	7.91	2.80	1,271	538	337	201	426	0.017	34	30	216	134	0.72
43.	8.33	2.31	2,388	900	537	363	302	0.041	62	38	486	119	1.09
44.	8.46	1.27	442	215	169	46	27	0.262	47	33	17	122	0.21
45.	8.46	1.10	4.93	1.88	96	92	9	0.012	37	36	51	143	1.47
46.	8.34	0.60	738	242	146	96	13	0.024	22	38	23	143	1.47
47.	8.34	0.60	738	242	146	96	13	0.023	15	33	11	86	0.56
48.	8.32	0.54	717	188	129	59	18	0.013	7	25	27	64	0.54
49.	7.38	0.61	469	188	56	62	9	0.067	2	4	26	263	0.33
50.	6.91	0.07	122	56	34	22	9	0.014	2	3	3	17	0.06
51.	7.19	0.06	556	81	51	30	9	0.018	2	3	6	14	0.08
52.	7.86	1.45	539	242	145	97	169	0.023	6	28	148	112	0.33
53.	7.98	0.08	162	54	32	22	36	0.028	1	3	721	71	0.16
54.	7.52	2.03	662	450	253	197	107	0.027	40	37	45	261	1.91
55.	7.64	0.98	604	337	225	212	36	0.028	28	32	12	134	1.31
56.	8.39	1.00	503	269	169	100	98	0.014	28	34	34	142	1.32
57.	7.65	1.59	708	323	108	143	44	0.012	13	31	145	90	0.69

All values are Expressed in mg/L Except pH and E.C.

** E.C. is measured in m. mhos/cm.

* Ca^{2+} and Mg^{2+} are expressed as $CaCO_3$.

Table 4.2 : Typical Analysis of Textile and Dyeing Industry Effluents in Coimbatore

Sl. No.	*Parameter*	*Textile effluent*	*Dyeing effluent*
1.	pH	10.4	12.9
2.	Electrical conductivity*	30.6	28.2
3.	Total suspended solids	560	598
4.	Total dissolved solids	8,218	7,613
5.	Hardness as $CaCO_3$	45	28
6.	Calcium as $CaCO_3$	20	11
7.	Magnesium as $CaCO_3$	25	17
8.	Chloride	2,552	1,118
9.	Nitrite	0.15	0.28
10.	Nitrate	49	68
11.	Phosphate	43	53
12.	Sulphate	890	836
13.	Alkalinity	1,200	2,325
14.	Fluoride	5.8	3,800
15.	Sodium	3,620	3,800
16.	Zinc	1.20	1.01
17.	Lead	0.48	0.59
18.	Copper	0.56	0.49
19.	Cadmium	0.05	0.05
20.	Iron	3.20	2.74

* inn m.mhos.

(*c*) the effluents of dyeing and textile mills are highly alkaline.

(*d*) the sodium content in the effluents of textile and dying mills is very high.

In the stage of textile mill operations (Nemerow, 1978) disizing, scouring, bleaching, mercerising and dyeing, large quantities of waste water are produced. This waste water contains high concentrations of sodium, other solids, high biological oxygen demand and alkalinity. The effluents are let into drainage canals which are not seepage proof. So these effluents seep through the soil and contaminate the groundwater. High concentrations of sodium in the effluent will exchange the calcium and magnesium in the soil and thus release calcium and magnesium into the groundwater. Thus it is evident that the high hardness of groundwater is only due to the seepage of the effluents from textile industries and dyeing units. It can be predicted that there is no possibility of reduction in the hardness as the industrial effluents are flowing in the drainage canals continuously throughout the year. The hardness in the well waters decreasing only during rainy season. During this season only, the people of these areas use this groundwater for domestic purposes other than drinking. Since only limited quantity of water is being withdrawn, there is no scope for the hardness to be completely removed even in the future.

High concentrations of nitrite are found in some wells in the city and also in the surrounding villages which are in close proximity to drainage canals through which sewage and sullage flows. The concentrations of nitrate and phosphate are also high except in Siruvani reservoirs. This may be attributed to the seepage of the sewage through the soils. The fluoride concentrations are slightly higher in a few places in the surrounding villages of the city whereas they are normal in all other samples. Sulphate concentrations are found to be very high in the wells in the areas whereas many textile mills are located. Cadmium is not in the detectable range in any of the well waters. Other metals like zinc, lead, copper and iron are slightly higher in the well waters where textile mills are located compared to the other wells. However, their concentrations are within the permissible limits (World Health Organisation, 1984).

From the above discussion, it is clear that the hardness in groundwaters of Coimbatore city is only due to the seepage of effluents of textile industries and dyeing units containing high concentrations of sodium.

Acknowledgements

The authors express their grateful thanks to Prof. M.N. Sastry, School of Chemistry, Andhra University and Prof. M.V. Venkata Rao, Department of Civil Engineering, Andhra University for their useful discussions and encouragement. The authors are also thankful to the Department of Environment, New Delhi for financial assistance in the form of a research project to the first author.

References

Franson, M.A.H., (Ed.), American Public Health Association, American Water Works Association and Water Pollution Control Federation, (1985) Standard methods for the examination of water and waste water, Washington. D.C.

Golterman, H.L., Clyno, R.S. and Ohnstad, M.A.M., (1978) IBP Hand book of methods of physical and chemical analysis of fresh waters, Blackwell Scientific Publications, Oxford.

Jan, T.K. and Young, D.R., (1978). Determination of microgram amounts of some transition metals in sea water by methyl isobutyl ketone-nitric acid successive extraction and flameless atomic absorption spectrophotometry. Anal. Chem., v. 50, pp. 1250.

Nemerow, N.L., (1978). Industrial water pollution–Origins, Characteristic & Treatment Addition–Wesley Publishing Co., U.S.A.

Vogel, A.I., (1978). Quantitative inorganic analysis. Longman Group Ltd., England.

W.H.O., (1984). Guidelines for drinking water quality (Vol. 2). World Health Organisation, Geneva.

5

Study on the Suitability of Groundwater for Drinking and Irrigation from a Coastal Sandstone Belt, Guntur District, Andhra Pradesh

P. SANKARA PITCHAIAH
AND
D. SRINIVASU

Introduction

Groundwater, which is always in motion (except under some conditions) interacts with the subsurface environment. During this movement, it dissolves, transports and deposits mineral matter. The health and economic prosperity of a community rest on the supply of clean water to an individual. In India, it is estimated that 73 million work days are lost every year due to water related diseases. The cost of treating them and the loss in production amount to Rs. 600 crores (India's Environment, 1982).

The quality of groundwater varies from place to place and stratum to stratum. Further, the quality of water depends on the specific need. Hence, the assessment of the quality of water in any area is a prerequisite. The present study is aimed at assessing the groundwater quality of the coastal sandstone belt of Guntur district for purpose of drinking and irrigation.

The area under investigation is lying at latitudes 16°10′–16°18′N and longitudes 80°28′–80°37′E (Fig. 5.1). Semi-arid type of climate is prevailing in the area. The average annual rainfall is 874 mm.

The aquifers are composed of sandstones and the sandstones are confined to a narrow zone of 50 sq.km. Analysis of pumping tests conducted in a few localities (State Ground Water Board, Unpubl. Report, 1990) indicates that the transmissivity of the aquifers ranges from 777.4 to 945 m^2/day. The specific capacity of the aquifer varies from 107.4 to 123.6 m^2/day. The recharge of water in this area is mainly due to precipitation. Most of the wells in the area tapping water form the shallow aquifer zone. The wells range in depth from 9 to 20 m.

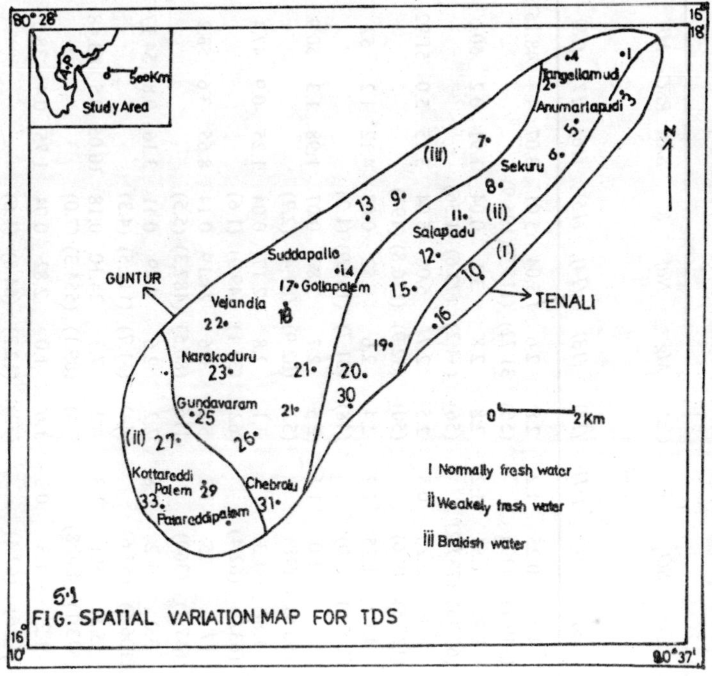

FIG. 5.1 SPATIAL VARIATION MAP FOR TDS

Table 5.1 : Chemical Analysis of the Water Samples from Coastal Sand Stone Belt (October, 1989)

Sample No.	Temp °C	pH	E.C. mmohs/cm	TDS mg/l	Total Hardness	CO_3^{--}	HCO_3^-	Cl^-	SO_4^{--}	F^-	Ca^{++}	Mg^{++}	Na^+	K^+	SAR	RSC	MH	%Na
(1)	(2)	(3)	(4)	(5)	(6)	(7)	(8)	(9)	(10)	(11)	(12)	(13)	(14)	(15)	(16)	(17)	(18)	(19)
1.	26.3	8.08	1090	697	5.4 (87.7)	–	9.8 (597.8)	4.2 (149.1)	0.25 (12)	1.2	2.8 (56)	2.6 31.72)	5.04 (115.9)	5.01 (195.9)	3.07	4.4	48.15	65.05
2.	26.9	7.40	1202	769	5.6 (90.2)	T.R.	5.4 (329.4)	1.9 (67.45)	1.54 (73.92)	1.2	2.8 (56)	2.8 (34.2)	3.2 (73.6)	0.04 (1.56)	1.91	0.2	50.0	36.65
3.	26.9	7.09	452	289	5.2 (82.9)	–	10.2 (622.2)	5.0 (177.5)	0.20 (9.6)	0.8	2.5 (50)	2.7 (32.9)	5.08 (116.8)	5.04 (197)	3.15	5.0	51.92	66.05
4.	26.8	8.01	1320	844	5.0 (79.7)	–	6.2 (378.2)	4.1 (145.5)	1.25 (60)	1.2	2.4 (48)	2.6 (31.7)	6.52 (149.9)	0.03 (1.2)	4.12	1.2	52	56.71
5.	27.1	7.43	798	511	5.3 (84.9)	–	6.6 (402.6)	1.6 (56.8)	1.0 (48)	1.1	2.6 (52)	2.7 (32.9)	3.80 (87.4)	0.07 (2.7)	1.98	1.3	50.9	42.20
6.	26.8	7.56	786	503	5.9 (96.1)	–	5.0 (305)	1.8 (63.9)	1.3 (62.4)	1 1	3.1 (62)	2.8 (34.1)	2.17 (49.9)	0.04 (1.6)	1.25	-0.9	47.4	27.25
7.	27.0	7.39	2760	1766	12.4 (196.5)	–	8.8 (536.8)	17.4 (617.7)	7.5 (360)	1.2	5.8 (116)	6.6 (80.5)	21.19 (487.3)	0.14 (5.5)	8.65	-3.6	53.2	63.24
8.	27.2	8.02	1220	780	4.8 (75.7)	T.R.	5.6 (341.6)	3.0 (106.5)	1.2 (57.6)	0.8	2.2 (44)	2.6 (31.7)	4.89 (112.5)	0.11 (4.3)	3.16	0.8	54.17	51.02
9.	26.5	8.24	2470	1580	11.5 (173.1)	T.R.	11.0 (671)	15.3 (543.1)	9.1 (436.8)	1.8	4.2 (84)	7.3 (89.1)	24.10 (554.3)	0.18 (7.0)	10.05	0.5	63.48	67.86
10.	26.9	8.27	534	342	2.6 (44.2)	T.R.	3.0 (183)	1.2 (42.6)	1.3 (62.4)	0.3	1.6 (32)	1.0 (12.2)	2.82 (64.8)	0.04 (1.6)	1.75	0.4	38.5	52.38

11.	26.5	7.98	1020	653	3.5 (59.0)	–	4.4 (268.4)	1.8 (63.9)	1.48 (71)	1.4	2.1 (42)	1.4 (17.0)	4.02 (92.5)	0.16 (6.2)	3.04	0.9	40.0	54.43
12.	26.2	7.92	1170	749	4.0 (64.4)	T.R.	3.9 (237.9)	2.2 (78.1)	1.34 (64.3)	1.6	2.0 (40)	2.0 (24.4)	3.4 (78.2)	0.04 (1.6)	2.40	-0.1	50.0	46.24
13.	27.0	7.78	3230	2067	31.0 (488.9)	T.R.	8.4 (512.4)	36.5 (1295.75)	22.44 (1077.1)	1.4	14.2 (284)	16.8 (204.9)	36.2 (832.6)	0.14 (5.5)	9.19	22.6	54.19	53.96
14.	27.2	7.04	3400	4224	38.5 (621.8)	–	10.2 (622.2)	43.4 (1540.7)	26.9 (1291.2)	1.2	19.5 (390)	19.0 (231.8)	40.76 (937.5)	1.28 (50.0)	9.12	28.3	49.3	52.19
15.	26.5	8.25	1190	762	4.0 (63.6)	T.R.	4.6 (280.6)	1.9 (67.4)	0.68 (32.6)	0.9	1.9 (38)	2.1 (25.6)	3.1 (71.3)	0.08 (3.1)	2.19	0.6	52.5	44.29
16.	26.5	7.66	570	326	3.0 (50.6)	–	3.8 (231.8)	1.6 (56.8)	1.8 (86.4)	1.6	1.8 (36)	1.2 (14.6)	4.07 (93.6)	0.11 (4.3)	3.38	0.8	40.0	58.21
17.	27.2	7.56	3510	2502	12.4 (178.8)	T.R.	9.6 (585.6)	15.2 (539.6)	9.1 (436.8)	1.0	4.0 (80)	8.1 (98.8)	22.04 (506.9)	0.15 (5.8)	8.96	-2.5	66.9	64.71
18.	27.0	7.14	3180	2035	13.3 (193.5)	–	12.4 (756.4)	14.0 (497)	12.1 (580.8)	0.6	4.0 (80)	9.3 (113.5)	24.99 (574.8)	0.19 (7.4)	9.41	-0.9	69.9	64.71
19.	26.0	7.86	888	568	6.3 (107.3)	T.R.	5.6 (341.6)	2.6 (92.3)	1.2 (57.6)	0.7	3.9 (78)	2.4 (29.3)	3.04 (69.9)	0.04 (1.6)	11.75	-0.7	38.1	32.83
20.	26.5	8.10	1280	819	4.2 (71.5)	–	4.2 (256.2)	1.9 (67.4)	2.6 (124.8)	1.2	2.6 (52)	1.6 (19.5)	4.32 (99.4)	0.18 (7.0)	2.98	–	38.09	51.72
21.	27.2	7.38	2760	1766	14.6 (235.8)	T.R.	8.6 (524.6)	16.2 (575.1)	5.1 (244.8)	1.1	7.4 (148)	7.2 (87.8)	15.21 (349.8)	0.13 (5.1)	5.59	-6.0	49.3	51.23
22.	26.6	7.92	3250	2080	15.8 (253.6)	–	8.4 (512.4)	17.6 (624.8)	6.09 (292.3)	1.4	7.8 (156)	8.0 (97.6)	16.21 (372.8)	0.08 (3.1)	5.76	-7.4	50.63	50.76
23.	26.8	7.82	2480	3507	20.0 (318.9)	–	10.6 (646.6)	20.2 (717.1)	7.64 (366.7)	1.6	9.6 (192)	10.4 (126.9)	18.4 (423.2)	0.04 (1.6)	5.82	-9.4	52.0	47.97

(1)	(2)	(3)	(4)	(5)	(6)	(7)	(8)	(9)	(10)	(11)	(12)	(13)	(14)	(15)	(16)	(17)	(18)	(19)
24.	27.0	7.46	3060	999	5.4 (86.2)	–	9.6 (585.6)	4.6 (163.3)	2.45 (117.6)	1.0	2.6 (52)	2.8 (34.2)	5.98 (135.5)	5.27 (206.0)	2.75	4.2	48.1	67.39
25.	26.8	7.75	2850	2464	11.6 (183.6)	–	8.9 (524.9)	8.9 (315.9)	9.14 (438.7)	1.6	5.4 (108)	6.2 (75.6)	14.2 (326.6)	1.14 (44.6)	5.89	-2.7	53.45	56.94
26.	26.6	8.25	2030	1299	6.0 (97.4)	T.R.	5.4 (329.4)	11.12 (394.7)	2.34 (112.3)	1.8	3.1 (62)	2.9 (35.4)	12.8 (294.4)	0.06 (2.3)	7.39	-0.6	48.33	33.13
27.	27.2	8.24	1094	700	4.6 (74.8)	–	4.3 (262.3)	1.8 (63.9)	0.78 (37.4)	1.0	2.4 (48)	2.2 (26.8)	2.24 (51.5)	0.04 (1.5)	1.47	-0.3	48.33	33.13
28.	27.1	7.80	2440	3289	11.5 (190.2)	–	10.4 (634.4)	12.4 (440.2)	9.22 (442.5)	1.4	6.4 (128)	5.1 (62.2)	16.4 (377.2)	4.12 (161.1)	6.39	-1.1	44.35	64.08
29.	26.9	7.97	1230	787	3.9 (61.6)	T.R.	4.1 (250.1)	2.2 (78.1)	2.0 (96)	1.0	1.8 (36)	2.1 (25.6)	4.32 (99.4)	0.08 (3.1)	3.09	0.2	53.85	53.012
30.	27.1	8.06	1270	813	4.3 (69.6)	T.R.	5.2 (317.2)	3.0 (106.5)	0.04 (1.92)	1.0	2.2 (44)	2.1 (25.6)	3.82 (87.8)	0.12 (4.7)	2.60	0.9	48.84	47.81
31.	27.0	7.90	2220	1421	19.2 (91.3)	T.R.	2.3 (140.3)	2.6 (92.3)	3.4 (163.2)	1.0	3.1 (62)	2.4 (29.3)	2.17 (49.9)	0.08 (3.1)	0.7	-6.9	12.5	10.48
32.	26.9	7.42	1250	800	2.9 (50.2)	–	3.2 (195.2)	1.2 (42.6)	10.2 (489.6)	0.8	1.9 (38)	1.0 (912.2)	11.41 (262.4)	0.11 (4.3)	9.47	0.3	34.48	97.89
33.	27.0	7.68	1100	704	4.0 (63.6)	–	3.9 (237.9)	2.2 (78.1)	0.73 (35.0)	1.6	1.9 (38)	2.1 (25.6)	2.83 (65.1)	0.03 (1.2)	2.00	-0.1	52.5	41.69

(Figures given in brackets are in mg/l); TR-trace; –Nil

Material and Methods

Groundwater samples were collected from 12 villages, namely, Tangellamudi, Anumarlapudi, Sekuru, Selapadu, Suddapalli, Gollapalem, Vejendla, Narakoduru, Gundavaram, Kottareddy Palem, Patareddy Palem and Chebrolu of Guntur district. Temperature, pH, electrical conductivity (EC), total dissolved solids (TDS), hardness, carbonates, bicarbonates, fluorides, sulphates, sodium and potassium were estimated. These chemical parameters were compared with the Indian Standard Specifications (1983) and National Academy of Sciences and National Academy of Engineering (1972) to assess their suitability for drinking (for domestic, livestock and poultry).

Water quality requirement for purpose of irrigation were studied using various empirical indices (US Department of Interior (1968), Doorenbas and Pruitt (1975), Sawyers and Westcot (1976) and Shainberg and Oster (1978). These indices were represented in irrigation water quality diagrams and they were compared with the irrigation water classification (US Salinity Laboratory, 1954) and Wilcox Classification (Wilcox, 1955). Bicarbonate hazard-BH (Eaton, 1950) and magnesium hazard-MH (Szablocs and Darab, 1964) were also calculated to quantify the effects of bicarbonate and magnesium.

Results and Discussion

The chemical parameters analysed in the groundwater of the sandstone belt are presented (in meq/l) in Table 5.1 and their suitability for drinking and irrigation is discussed.

Drinking

1. Domestic use

(A) *Temperature* : The minimum and maximum temperature of the groundwater in the study area are 26°C and 27.2°C respectively. In comparison to the standards suggested by Indian Standard Specifications for Drinking Water (1983), these values fall within the normal range.

(B) *pH* : pH values in the groundwater from the sandstone belt vary from 7.04 to 8.27. These values indicate slight alkaline nature.

(C) *TDS* : TDS values range from 289.3 to 4224 mg/l. Groundwater in the area is classified depending on the TDS values (Ranga Rao, 1987, unpubl. thesis). Based on this, spatial variation map is prepared to study the quality of the groundwater in the study area (Fig. 5.1).

Normally fresh water is found around Anumarlapudi and along the eastern part of the sandstone belt. Weakly freshwater is identified around

Tangellamudi, Sekuru, Selapadu, Patareddy Palem and Kotareddy Palem. Brackish water is distributed around Chebrolu, Gundavaram, Narakoduru, Vejendla, Gollapalem and Suddapalli villages.

(D) *Hardness* : The hardness of the water samples collected in the study area ranges from 44.2 to 621.8 mg/l. These samples can be classified into three categories depending on the hardness values (Sawyer and McCarty, 1967). They are soft, moderately hard, hard and very hard waters. Soft water is found around Selapadu, Kottareddy Palem and Patareddy Palem. Moderately hard water is existed around Tangellamudi, Anumarlapudi, Sekuru and Chebrolu. Hard water is found around Vajendla, Gollapalem and Gundavaram and very hard water is found around Suddapalli and Narakoduru.

(E) *Chloride* : Chloride ion ranges from 42.6 mg/l to 1540 mg/l. One third of the water samples from the study area, have chloride concentration more than the desirable (250 mg/l). The northern and the southern part of the sandstone belt is characterised with these higher chloride concentrations. It is suggested to utilise this water when there is no alternative.

(F) *Fluoride* : Fluoride ion ranges from 0.3 mg/l to 1.8mg/l. These values indicate that the ion is within the permissible limits and there is no problem in using this water in the study area for drinking purpose.

Sulphate : In the study area sulphate ion ranges from 2 to 1291 mg/l. High concentrations are recorded all along the north and northwestern parts of the sandstone belt.

2. Livestock and Poultry

Groundwater in the study area is classified into three categories, namely, excellent, very satisfactory and satisfactory following the classification given by National Academy of Sciences and National Academy of Engineering (1972).

Excellent water is found around Tangellamudi, Anumarlapudi, Sekuru, Selapadu, Kottareddy Palem and Patareddy Palem. This water is excellent for all classes of livestock and poultry. Very satisfactory water is present around Gollapalem, Vejendla, Gundavaram and Chebrolu. This water may cause temporary and mild diarrhoea in livestock and water droppings in poultry, which are not accustomed to it. Satisfactory water is found around Suddapalli and Narakoduru villages. This water may cause temporary diarrhoea or be refused at first by animals not accustomed to it. It is a poor quality water for poultry and it may cause water faeces, increased mortality and decreased growth, especially in turkeys.

Irrigation Use

(A) *Total Salt Concentration Hazard* : Based on the USSL clasification (United States Salinity Laboratory, 1954) it is concluded that the groundwater from the sandstone belt is highly saline to very highly saline in nature. This water is not suitable for all crops and it is suggested to select salt tolerant varieties.

(B) *Sodium Hazard* : With reference to Wilcox classification (1955) most of the water samples from the coastal sandstones can be classified into low and medium sodium water classes.

Wilcox diagram is prepared (Fig. 5.2) to classify the groundwater from the study area for irrigation purpose on the basis of per cent sodium and specific conductivities. Figure 5.2 supports that the majority of the samples belong to two classes–good to permissible and doubtful to unsuitable.

Spatial variation map is prepared for the Wilcox diagram (Fig. 5.3) and found that the excellent to good waters are present near Anumarlapudi and along the eastern part of the sandstone belt. Good to permissible water is located around Sekuru, Selapadu and Kottareddy Palem villages. Tangellamudi and Patareddy Palem villages are characterised by the presence of permissible to doubtful water. Doubtful to unsuitable water is found around Narakoduru, Gundavaram and Chebrolu and the unsuitable water is found around Suddapalli, Gollapalem and Vejendla villages.

Sodium hazard is also studied in comparison to US Department of Agriculture classification (1954). According to this, half of the groundwater samples are good for irrigation and the remaining are moderate quality water for irrigation (Srinivas, Unpubl. thesis, 1990).

(C) *Bicarbonate Hazard* : The Residual Sodium Carbonate (RSC) values of the groundwater samples (Table 5.1) support that most of the groundwater (82%) in the study area is free from bicarbonate hazard.

(D) *Magnesium Hazard* : Most of the groundwater samples from the sandstone belt have MH values less than 50. It indicates that this water is free from magnesium hazard.

Acknowledgements

The authors are grateful to Late Prof. Y.L. Dora, Head of the Department, Geology, for the facilities provided in the department to carry out this work.

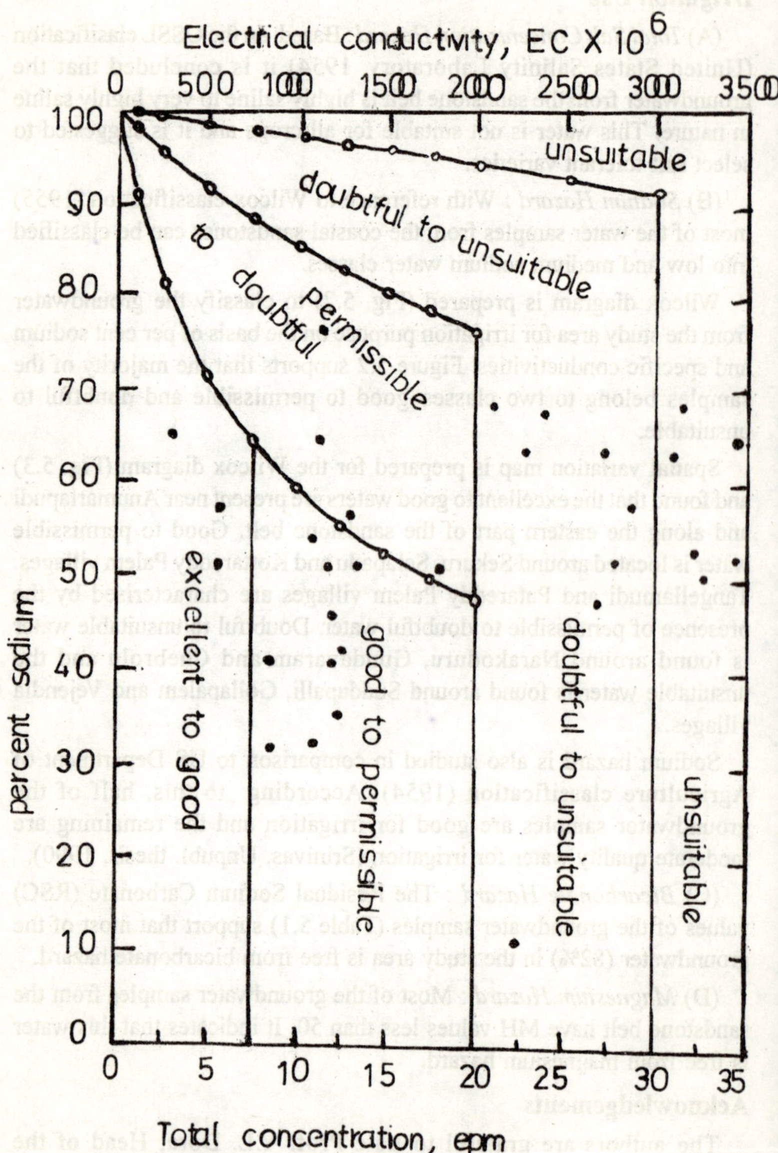

FIG 5·2 WILCOX DIAGRAM

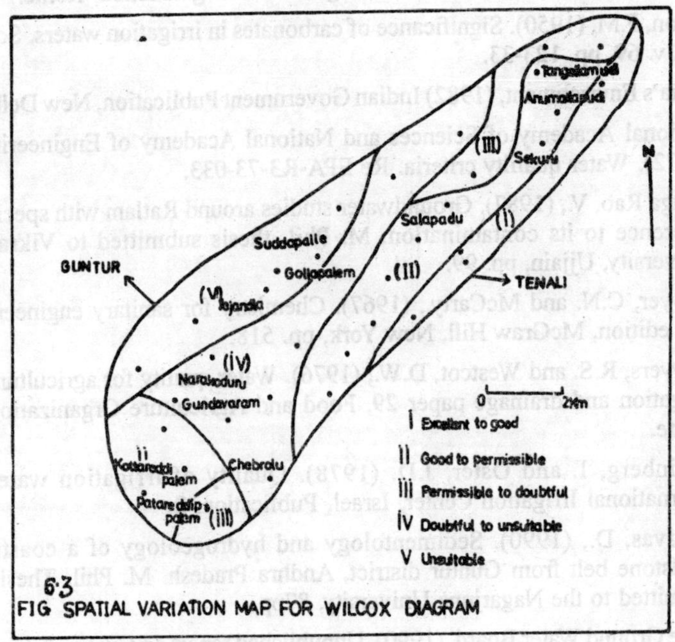

6.3
FIG SPATIAL VARIATION MAP FOR WILCOX DIAGRAM

References

Doorenbos, J. and Pruitt, W.O., (1975) Crop requirements. Irrigation and drainage paper 24, Food and Agriculture Organization, Rome.

Eaton, F.M, (1950). Significance of carbonates in irrigation waters. Soil. Sci. v. 69, pp. 123-33.

India's Environment, (1982) Indian Government Publication, New Delhi.

National Academy of Sciences and National Academy of Engineering (1972). Water qualilty criteria. Re EPA-R3-73-033.

Ranga Rao, V., (1987). Groundwater studies around Ratlam with special reference to its contamination, M. Phil. thesis submitted to Vikram University, Ujjain, pp. 99.

Sawyer, C.N. and McCarty., (1967). Chemistry for sanitary engineers, 2nd edition, McGraw Hill, New York, pp. 518.

Sawyers, R.S. and Westcot, D.W., (1976). Water quality for agriculture. Irrigation and drainage paper 29, Food and Agriculture Organization, Rome.

Shainberg, I. and Oster, J.D. (1978). Quality of irrigation water, International Irrigation Center, Israel, Publication, 2.

Srinivas, D., (1990). Sedimentology and hydrogeology of a coastal sandstone belt from Guntur district, Andhra Pradesh. M. Phil. Thesis, submitted to the Nagarjuna University, 87pp.

State Ground Water Board, (1990). Unpublished report on Guntur district.

Szablocs, I. and Darab, C., (1964). The influence of analysis, India and Engineering Chemistry. V. 15, pp. 394.

U.S. Department of Interior (1968). Report of the committee on water quality criteria, U.S. Dept. of Interior, Washington, D.C.

U.S. Salinity Laboratory (1954). Diagnosis and improvement of saline and alkali soils. U.S. Dept. of Agriculture. Hand Book. 60. pp. 160.

Wilcox, L.V., (1955). Classification and use of irrigation waters, U.S. Dept. of Agriculture, Circ. 968, Washington, D.C. pp. 19.

6

Hydrogeochemical Studies of Groundwater in Mamundiyar Basin, Tamil Nadu, India

P.R. Srinivasan, N. Dhakshinamoorthy and
A. Balasubramanian

Introduction

The Mamundiyar basin (latitudes 10° 25′ to 10°40′N and longitudes 78°10′ to 78°30′E) forms a part of Cauvery basin, Tamil Nadu, India. Extending over an area of approximately 620 km, the basin lies mostly in the Manapparai taluk of Tiruchirapalli and partly in the Viralimalai taluk of Pudukottai (Fig. 6.1).

The Mamundiyar river originates from the Irungadu group of hills and flows through the plains almost parallel to the National Highway 45 and finally joins the Ariyavur river about 10 km. NW of Maravanur village. Ponnanaiyar and Sanngai rivers are the major tributaries of Mamundiyar.

The area gets most of its rainfall from the NE monsoon and the average annual rainfall ranges between 75 and 85cm. The groundwater occurs under phreatic conditions and the water table fluctuation ranges between 3 and 7m in the basin area. The depth to the water table varies from 2 to 18 m bgl. The basin comprises of hard crystallines, namely, biotite gneiss, granitic gneiss, quartzites and intrusions of quartz veins, pegmatites and basic rocks like anorthosites of Archaean age. The soil is generally of red sandy type. Black cotton soil is noted in some areas of the eastern part of the basin.

Surface water facility being scanty, groundwater is the mainstay for the irrigational development. An attempt has been made in this paper to assess the qualitative characteristics of groundwater.

Materials and Methods

Representative water samples have been collected and analyzed using the standard procedures of water analysis (ISI, 1964). Following the methods listed by Richards (1954) the accuracy and precision of the analyses have been checked up. The chemical data relating to 42 ground water samples have been processed and analysed using HYCH-BASIC

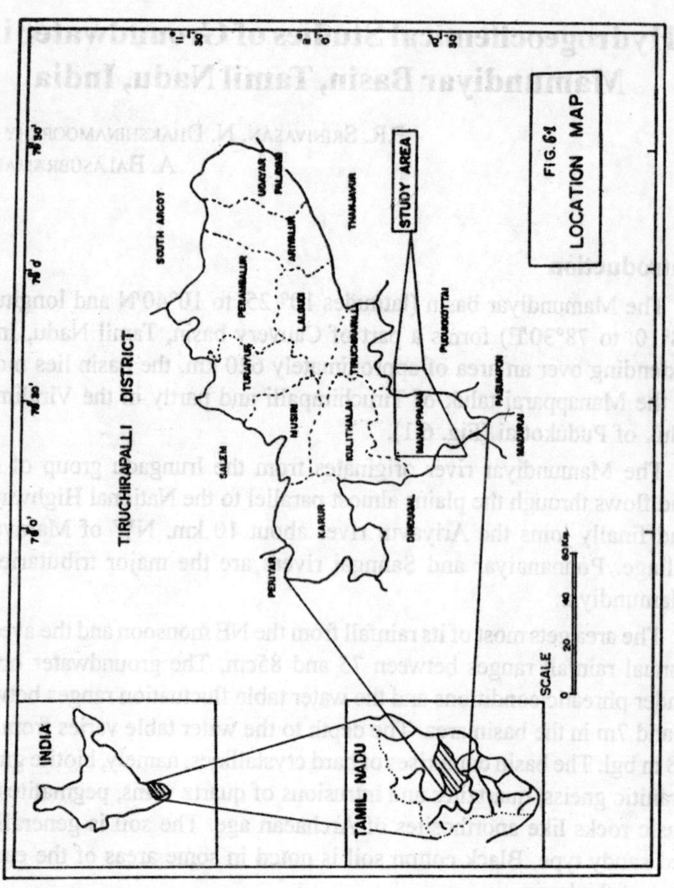
LOCATION MAP
FIG. 6.1
STUDY AREA
TIRUCHIRAPALLI DISTRICT
INDIA
TAMIL NADU
SCALE
0 20 40 50Km
SOUTH ARCOT
THANJAVUR
PUDUKOTTAI
SALEM
KARUR
PERIYAR
DINDIGAL
MADURAI
PERAMBALUR
ARIYALUR
LALGUDI
MUSIRI
KULITHALAI
MANAPPARAI
TURAIYUR
UDAIYAR PALAYAM
78°0'
78°30'
79°0'
79°30'

Computer programme developed by Balasubramanian *et al.* (1991). In the programme groundwater classification is attempted using the criteria of Handa (1964), Piper (1944), Stuyfzand (1989) and USSL (1955) schemes. The Index of Base Exchange (IBE) and groundwater types are also computed using Schoeller's procedures (1965, 1967).

Ryzner (1944) proposed the Corrosivity Ratio (CR) as an index to study the corrosive tendencies of flowing water in a metallic pipe. Balasubramanian *et al.* (1985) have used the above index to evaluate the corrosive nature of groundwaters of Tambraparni river basin, Tamil Nadu. The corrosivity ratio of groundwaters of Mamundiyar basin have been computed using,

$$\text{Corrosivity Ratio (C.R.)} = \frac{(Cl/35.5) + SO_4/48.0)}{(CO_3 + HCO_3)/50.0)}$$

All values are in ppm

Gibbs (1970) proposed plots making use of the chloroalkaline indices for inferring the mechanisms controlling the groundwater chemistry. However, in this present work instead of making use of the graphical procedure, numerical steps have been included in the programme. A sample output showing all the results obtained for one observation well has been given (Fig. 6.2).

Principal Component Analysis (PCA) is a simple mathematical reduction of the input data matrix, without any elaborate assumptions (Anderson 1958), Morrison (1964) and Wilks (1963). Detailed mathematical descriptions of PCA could be seen in Davis (1974). Identification and assessment of dominant ions have been made, using the loadings matrix and the component scores of this analysis.

A set of six spatial variation maps showing TDS concentration, ground water hardness, groundwater salinity and sodium hazards, groundwater types, Corrosivity ratio of groundwaters, and Hydrogeo-chemical model has been prepared to get an integrated picture of the regional quality behaviour of groundwater in the basin.

Results and Conclusions

The different quality parameters of groundwaters of Mamundiyar basin have been presented in Table-6.1. The results obtained and conclusions derived are summarised as follows.

1. Figure 6.3 shows the TDS concentrations of groundwater of this basin. The basin can be divided into three zones (TDS less than 500, 500-1000 and greater than 1000 ppm). This zonation has been done

FIG. 6.2

```
SAMPLE CODE=21   NAMMAMPATTI
------------------------------------------------------------------------
EC(mmhos)  = 1350           TDS (ppm)         = 864
pH         = 8.8            ORP               = 0
DO         = 0              Temp.(centig)     = 30
------------------------------------------------------------------------
Conc/Ion    Ca     Mg    Na+K   HCO3    CO3    Cl     NO3    SO4
------------------------------------------------------------------------
ppm        58.0   30.0  196.0  372.0   96.0  106.0   61.0   41.0
epm         2.9    2.5    8.5    6.1    3.2    3.0    1.0    0.9
%          20.8   17.8   61.4   65.8    3.2   21.2    7.0    6.0
------------------------------------------------------------------------
Sodium Absorption Ratio = 5.204105
Residual Sodium Carbonate= 3.93342
Non-carbonate Hardness  =-196.671
Permeability Index(Doneen)= 79.1585
IONIC STRENGTH =  0.0187        CORROSIVITY RATIO =  0.4103
INDICES OF BASE EXCHANGE =   -1.8502   -0.4969

          CaCO3 SATURATION INDICES :
Equilibrium Ca method=  0.7526      Equilibrium pH method=  1.8465
GIBB'S PLOT : MECHANISM CONTROLLING THE CHEMISTRY =  ROCK INTERACTION
------------------------------------------------------------------------
          HANDA'S CLASSIFICATION :
Hardness        =B2   Temporary
Salinity        =C3   Moderate
Sodium hazard   =S1   Low
------------------------------------------------------------------------
SCHOELLER'S WATER TYPE (r=epm)
     III  Since    rCL > rSO4 > rCO3
------------------------------------------------------------------------
          PIPER'S HYDROGEOCHEMICAL FACIES:
Cations  = Na+K, Ca+Mg          Anions       = HCO3+CO3, Cl+SO4
SIGNIFICANT ENVIRONMENT : DISSOLUTION AND MIXING
------------------------------------------------------------------------
          STUYFZAND'S CLASSIFICATION:
WATER TYPE(Based on Cl) =F-Fresh
SUB-TYPE(Based on Alk) =ALK-HIGH
FACIES          = NA+K     HCO3
SIGNIFICANT ENVIRONMENT :
(+) Na+Mg SURPLUS INDICATES FRESHWATER INTRUSION-ANYTIME ANYWHERE
------------------------------------------------------------------------
          USSL CLASSIFICATION :
Salinity        =C3            Sodium hazard = S2
```

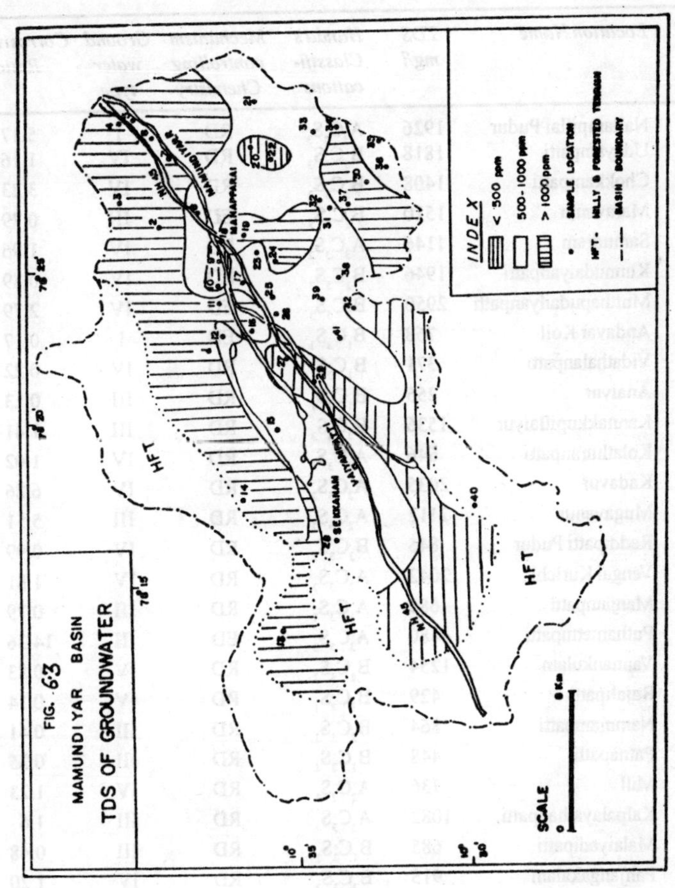

FIG. 6.3 MAMUNDIYAR BASIN
TDS OF GROUNDWATER

Table 6.1 : Different Chemical Characters of Ground Waters of Mamundiyar Basin

Well No.	*Location Name*	*TDS mg/l*	*Handa's Classifi-cations*	*Mechanism controlling Chemistry*	*Ground water type*	*Corrosivity Ratio*
1.	Nallampillai Pudur	1926	$A_3C_5S_3$	RD	I	5.17
2.	Udayampatti	1818	$B_3C_5S_3$	RD	IV	1.16
3.	Chokkampatti	1408	$B_3C_3S_3$	RD	IV	3.23
4.	Maravanur	1510	$B_2C_4S_3$	ED	III	0.79
5.	Samudram	1146	$A_3C_3S_2$	RD	IV	1.96
6.	Kunnudaiyanpatti	1946	$B_2C_5S_3$	ED	IV	0.89
7.	Mutthapudailyanpatti	2950	$B_3C_5S_3$	ED	IV	2.79
8.	Andavar Koil	768	$B_1C_3S_1$	ED	I	0.17
9.	Vidathalanpatti	371	$B_1C_2S_1$	RD	IV	0.22
10.	Anaiyur	255	$B_1C_3S_1$	RD	III	0.13
11.	Kannakkupillaiyur	1536	$B_3C_4S_3$	RD	III	1.41
12.	Kolathuranpatti	646	$A_3C_3S_1$	RD	IV	1.02
13.	Kadavur	2035	$A_2C_5S_2$	RD	IV	6.86
14.	Mugavanur	2112	$A_2C_5S_2$	RD	III	5.51
15.	Reddipatti Pudur	845	$B_2C_3S_1$	RD	IV	0.99
16.	Vengai Kurichi	2042	$A_2C_5S_3$	RD	IV	1.81
17.	Manjampatti	685	$A_1C_3S_1$	RD	III	0.79
18.	Puthamettupatti	4480	$A_3C_5S_3$	ED	III	14.36
19.	Vannankulam	1254	$B_2C_3S_2$	RD	IV	0.83
20.	Rajalipatti	429	$B_1C_2S_1$	RD	IV	0.14
21.	Nammampatti	864	$B_2C_3S_1$	RD	III	0.41
22.	Patnapatti	448	$B_1C_2S_1$	RD	III	0.35
23.	Mill	736	$A_2C_3S_1$	RD	IV	1.23
24.	Kalpalayathanpatti	1082	$A_2C_3S_1$	RD	III	1.6
25.	Malaiyadipatti	685	$B_1C_3S_1$	RD	III	0.48
26.	Panjangakalam	915	$B_3C_3S_2$	RD	IV	1.20
27.	Thomas Nagar	416	$B_1C_2S_1$	RD	III	0.24
28.	Sekkanam	704	$A_2C_3S_1$	RD	IV	1.2
29.	Mettukkadai	301	$B_2C_2S_1$	RD	III	0.42
30.	Thottiyapatti	1024	$A_3C_3S_1$	RD	IV	3.21
31.	Periyamanapatti	530	$B_1C_3S_1$	RD	IV	0.53
32.	Kilarampati	992	$A_3C_3S_1$	RD	IV	1.89
33.	Thengaithinnipatti	608	$B_1C_3S_1$	RD	I	0.43
34.	Vadakumuthapatti	1056	$B_2C_3S_2$	RD	III	0.49
35.	Kasavanur	851	$B_2C_3S_1$	RD	III	0.83
36.	Reddaipatti	749	$A_1C_3S_1$	RD	III	0.58

(Contd..)

Well No.	*Location Name*	*TDS mg/l*	*Handa's Classifi-cations*	*Mechanism controlling Chemistry*	*Ground water type*	*Corrosivity Ratio*
37.	Chinnamanapatti	3731	$A_3C_5S_3$	ED	IV	9.66
38.	Penna Kombu Natham	877	$A_1C_3S_1$	RD	I	0.60
39.	Pinnathur	614	$B_1C_3S_1$	RD	III	0.50
40.	Yelamanam	333	$A_1C_2S_1$	RD	III	0.84
41.	Murampatti	755	$A_1C_3S_1$	RD	IV	0.76
42.	Kavalkaranpatti	429	$B_1C_2S_1$	RD	III	0.22

RD–Rock Dominance ED–Evaporation Dominance

following Table 4.5 of Davis and De-Wiest (1966) to demarcate the suitable areas for groundwater for various uses like potability, irrigation and industry.

2. Figure 6.4 is the pictorial representation of groundwater hardness evaluated using Handa's classification in the basin. The basin could be delineated into six areas (A1, A2, A3 of Non-carbonate/Permanent hardness; B1, B2, B3 of Carbonate/Temporary hardness). The most undesirable type (A3) of groundwater is noticed in two pockets, one in the central part of the western side and another in the north eastern corner of the basin. For domestic and industrial development activities of the basin this figure could be conveniently used.

3. Groundwater salinity and sodium hazards (Handa, 1964) have been shown in Figure 6.5 in order to bring out the irrigational quality. Based on these parameters, the basin is divided into seven horizons (C_2S_1, C_3S_1, C_3S_2, C_3S_3, C_4S_3, C_5S_2 and C_5S_3). Amongst these : (1) C_5 type is unfit for agricultural use, (2) C_4 type can be used to some extent for raising salt tolerant crops like beetroot, (3) S_2 is suitable for permeable coarse grained soils and, (4) S_2 can be used only in areas with good drainage.

4. Figure 6.6 shows the groundwater types in the basin; type I: (CO_3 >Cl or SO_4), Type III: (Cl>SO_4>CO_3) and type IV: (Cl>SO_4>$CO_3$and Na> Mg>Ca). Thirty samples contribute to types III and IV of Schoeller (1965, 1967) and this indicates the greater residence time of waters in the subsurface and dominance of rock water interaction.

5. Groundwater is extracted from this basin for irrigation and domestic purposes. It is being transported by conventional metallic pipes. The suitability of metallic pipes for the transport of water depends upon the corrosivity ratio of water. A Corrosivity Ratio (C.R.) map (Fig. 6.7) has been drawn and the safe zone with CR<1.0 has been delineated. It may be added that in other areas (Fig. 6.7) PVC pipes will have to be used.

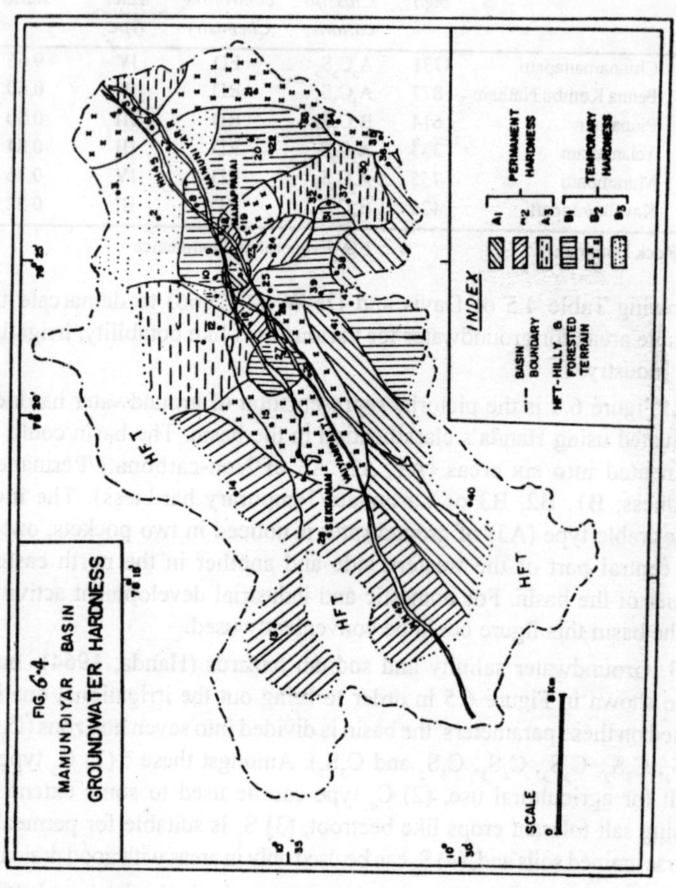

FIG. 6.4 MAMUNDIYAR BASIN GROUNDWATER HARDNESS

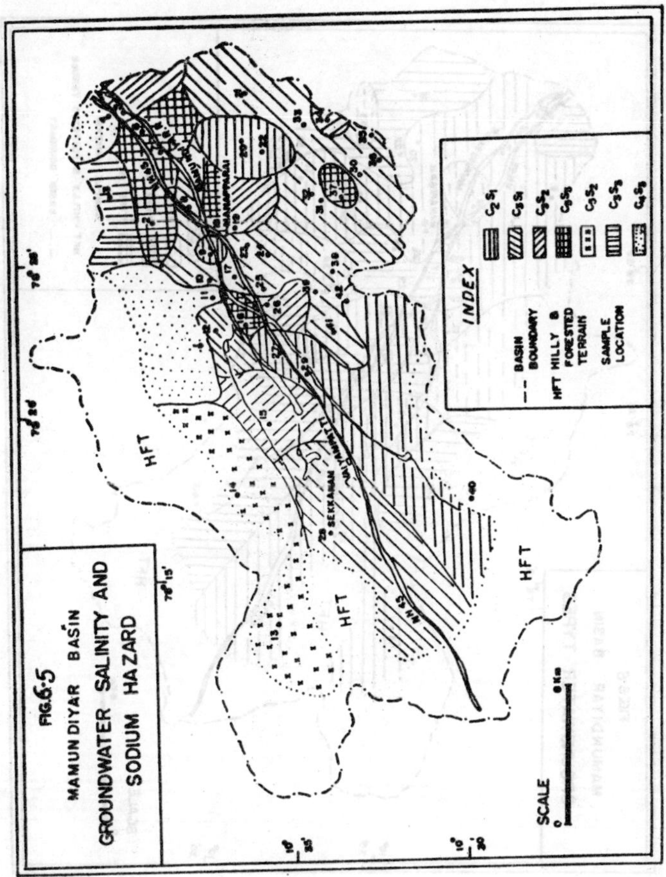
FIG.6.5
MAMUNDIYAR BASIN
GROUNDWATER SALINITY AND
SODIUM HAZARD
HFT
HFT
HFT
INDEX
BASIN BOUNDARY
HFT HILLY & FORESTED TERRAIN
SAMPLE LOCATION
SCALE
0
8 Km

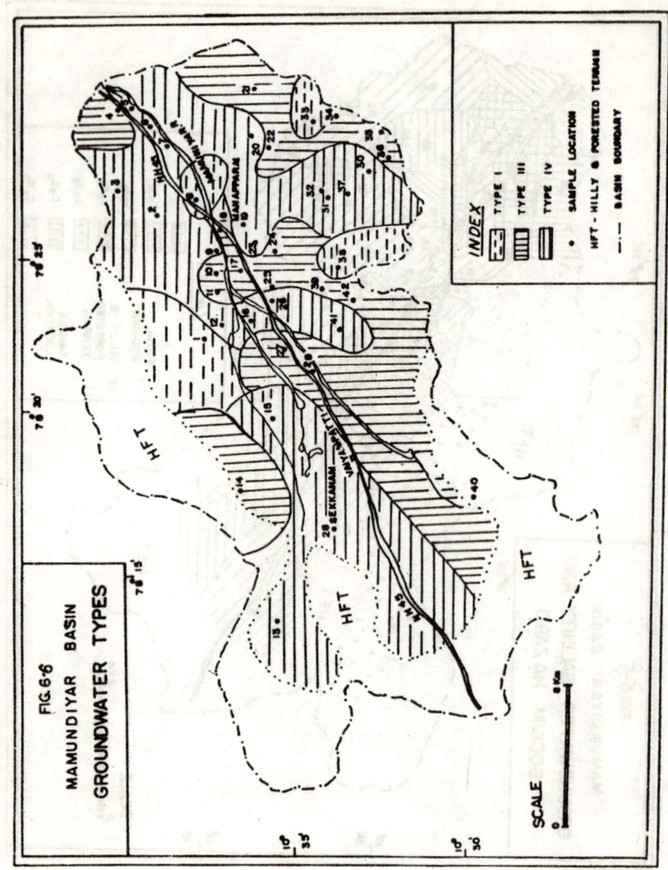

FIG. 6·6
MAMUNDIYAR BASIN
GROUNDWATER TYPES

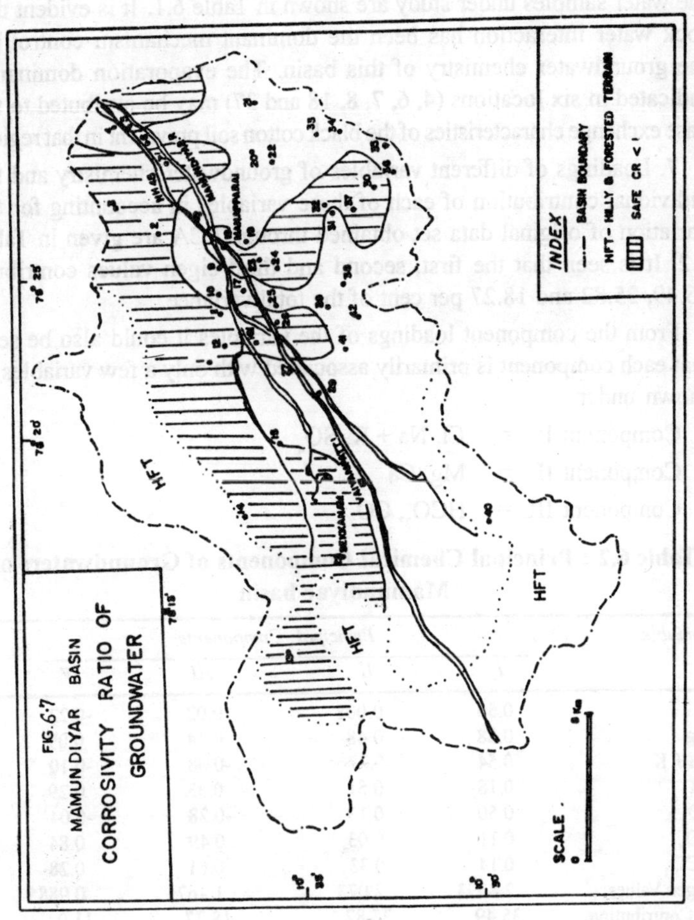

FIG. 6.7
MAMUNDIYAR BASIN
CORROSIVITY RATIO OF GROUNDWATER

6. The controlling factors of groundwater chemistry with reference to the water samples under study are shown in Table 6.1. It is evident that rock water interaction has been the dominant mechanism controlling the groundwater chemistry of this basin. The evaporation dominance indicated in six locations (4, 6, 7, 8, 18 and 37) may be attributed to the base exchange characteristics of the black cotton soil prevalent in that region.

7. Loadings of different variables of groundwater chemistry and the individual contribution of each of these variables in accounting for the variation of original data set obtained through PCA are given in Table 6.2. It is seen that the first, second and third eigen values contribute 35.49, 25.82 and 18.27 per cent of the total variance.

From the component loadings of the variables it could also be seen that each component is primarily associated with only a few variables as shown under.

Component I = Cl, Na + K, SO_4

Component II = Mg, Ca

Component III = HCO_3, CO_3

Table 6.2 : Principal Chemical Components of Groundwaters of Mamundiyar basin

Variables	*Principal Components*			
	I	*II*	*III*	*IV*
Cl	0.56	0.008	0.02	-0.22
Mg	0.28	0.48	0.34	0.09
Na + K	0.54	-0.27	-0.08	-0.10
Ca	0.18	0.51	0.33	0.29
SO_4	0.50	0.13	-0.28	-0.04
NO_3	0.11	0.03	-0.49	0.84
HCO_3	0.14	-0.33	0.61	0.28
Eigen Values	2.8393	2.073	1.467	0.9885
% Contribution of Eigen Values	35.49	25.82	18.27	11.0
Cumulative %	35.49	61.31	79.58	90.68

The dominance of Cl over the other ions here again can be taken as the evidence of the greater residence time of groundwaters (Freeze and Cherry, 1979).

8. Figure 6.8 has been brought out by mapping the positive scores of the first three Principal Components of individual samples. This hydro-geochemical facies model of the terrain gives the total picture of the hydrogeochemistry and can be used for any developmental planning.

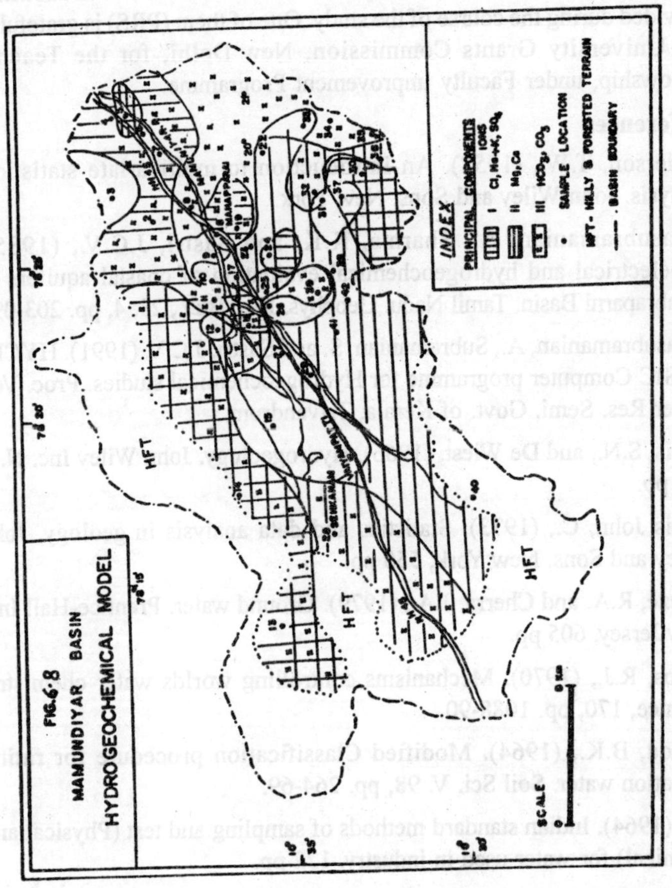

FIG. 6.8 MAMUNDIYAR BASIN HYDROGEOCHEMICAL MODEL

Acknowledgements

The authors thank their respective institutions for the facilities provided during the course of the study. One of them (PRS) is grateful to the University Grants Commission, New Delhi, for the Teacher Fellowship, under Faculty Improvement Programme.

References

Anderson, T.W., (1958). An Introduction to multivariate statistical analysis. John Wiley and Sons, New York.

Balasubramanian, A., Sharma, K.K. and Sastri, J.C.V., (1985). Geoelectrical and hydrogeochemical evaluation of coastal aquifers of Tambraparni Basin, Tamil Nadu. Geophys. Res. Bull., 23, 4, pp. 203-09.

Balasubramanian, A., Subramanian, S. and Sastri, J.C.V. (1991). HYCH-BASIC Computer programme for Hydrogeochemical studies. Proc. Vol. Water Res. Semi. Govt. of Kerala, Trivandrum.

Davis, S.N., and De Wiest, (1966). Hydrogeology. John Wiley Inc. N.Y. 463 pp.

Davis John, C., (1973). Statistics and data analysis in geology, John Wiley and Sons, New York, 550 pp.

Freeze, R.A. and Cherry, J.A., (1979). Ground water, Prentice-Hall Inc, New Jersey, 605 pp.

Gibbs, R.J., (1970). Mechanisms controlling worlds water chemistry, Science, 170, pp. 1088-90.

Handa, B.K., (1964). Modified Classification procedure for rating irrigation water. Soil Sci. V. 98, pp. 264-69.

ISI, (1964). Indian standard methods of sampling and test (Physical and chemical) for water used in industry. 122 pp.

Morrison, D.F., (1964). Multivariate statistical methods. McGraw Hill, New York.

Piper, A.M., (1944). A graphic procedure in the geochemical interpretation of water analysis. Am. Geophys. Union Trans., 25 pp. 914-23.

Richards, C.A., (1954). Diagnosis and improvement of saline and alkali soils. USDA hand book 60, 160 pp.

Rynzer, J.W., (1944). A new index for determining amount of Calcium carbonate scale formed by water. Jour A.W.W.A, V.30, No. 4 pp. 472.

Schoeller, (1965). Hydrodynamique dans Lekarat (Ecoulement et Emmagasinement) Actes Colloques Doubronik, V. I, pp. 3-20, AIHS et UNESCO.

Schoeller, H., (1967). Qualitative evaluation of groundwater Resources, (in Methods and techniques of groundwater investigation and development) Wat. Res. Series, 33, UNESCO, pp. 44-52.

Stuyfzand, P.S. (1989). A new hydrochemical classification of water types. IAMS Publ. 182, pp. 89-98.

USSL., (1955). Classification and use of Irrigation waters. U.S. Department of Agriculture, Circu. 969, Washington.

Wilks, S.S., (1963). Mathematical statistics. John Wiley and Sons, New York.

7

Hydrogeochemistry of Groundwater in a Coastal Sandstone Belt, Guntur District, Andhra Pradesh

P. SANKARA PITCHAIAH AND

D. SRINIVAS

Introduction

Coastal sandstones are exposed near Vadlamudi, in Guntur District (Fig. 7.1) (Lat. 16°10′-16°18′N and Long. 80°28′-80°37′E). Geologically, the area is characterised by the formation of the Upper Gondwana coastal sandstones (Geological Survey of India map, 1975) with different textures and structures. Semi-arid type of climate prevails in the area. Clays are present on the northern part of the sandstone belt whereas alluvial soils are seen on southern side.

Material and Methods

Representative groundwater samples collected from 33 wells in the sandstone belt and stored in one litre polythene bottles in the month of May, 1989 during summer season. Chemical parameters of the samples were analysed within 24 hours using standard laboratory techniques (Ramteke and Moghe, 1988).

In this study three methods Schoeller nomographs (Schoeller, 1959); Ionic ratios and Numerical approach (Haddad, 1981) and Durov's trilinear diagram (1958) were used to study the types of groundwater, hydrochemical facies and the chemical genesis of the groundwater. Haddad (1981) numerical approach was used to classify the groundwater. In this method, the relationship between major cations and anions was taken into consideration. The ratios were arranged as follows : Ca/Mg; Na + K/Mg; Na + K/Ca; Cl/Na + K; Cl/Mg; Cl/Ca; Cl + SO_4/HCO_3. From these arranged ratios three possibilities are existing; they are : ratios being less than one, equal to one and more than one and these ratios were represented with the notation 1, 2, 3 respectively. This approach is applicable for providing the rapid classification and evaluation of a larger number of samples (Eischamia and Hamdan, 1989).

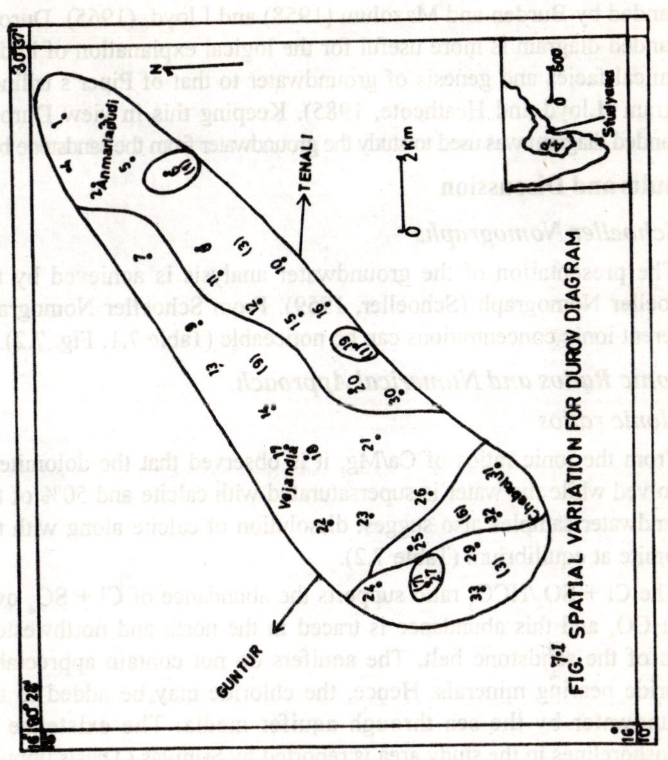

FIG. 7.1 SPATIAL VARIATION FOR DUROV DIAGRAM

The use of trilinear diagram to represent groundwater facies was first attempted by Hill (1940) and refined by Piper (1944). Durov (1948) devised an alternative diagram to that of piper (1944) which was further expanded by Burden and Mazolum (1958) and Lloyd, (1965). Durov's expanded diagram is more useful for the logical explanation of hydro-chemical facies and genesis of groundwater to that of Piper's trilinear diagram (Lloyd and Heathcote, 1985). Keeping this in view Durov's expanded diagram was used to study the groundwater from the sandstone belt.

Results and Discussion

1. Schoeller Nomographs

The presentation of the groundwater analysis is achieved by the Schoeller Nomograph (Schoeller, 1959). From Schoeller Nomograph different ionic concentrations can be noticeable (Table 7.1, Fig. 7.2).

2. Ionic Ratios and Numerical Approach

(a) Ionic ratios

From the ionic ratios of Ca/Mg, it is observed that the dolomite is dissolved while the water is supersaturated with calcite and 50% of the groundwater samples also suggest dissolution of calcite along with the dolomite at equilibrium (Table 7.2).

The Cl + SO_4/HCO_3 ratio supports the abundance of Cl + SO_4 over the HCO_3 and this abundance is traced in the north and northwestern parts of the sandstone belt. The aquifers do not contain appreciable chloride bearing minerals. Hence, the chloride may be added to the groundwater by the sea through aquifer media. The existence of paleoshorelines in the study area is reported by Srinivas (Thesis unpubl. 1990).

(b) Numerical approach

The major ion ratios and corresponding numerical symbols are shown in Table 7.2. Based on these symbols groundwater in the study area is classified into 20 groups. These groups indicate different combinations of the ionic concentrations in the groundwaters of the study area.

3. Durov's trilinear diagram

In the present study, an attempt is made to represent the groundwater analysis in Durov's expanded diagram (Fig. 7.3). It is observed that the groundwater samples are distributed among six fields (1, 2, 3, 4, 6, 7, and 9). However, the samples are distributed mainly between 3 and 9 fields. Following Burden and Mazloum (1958) and Lloyd (1965) various

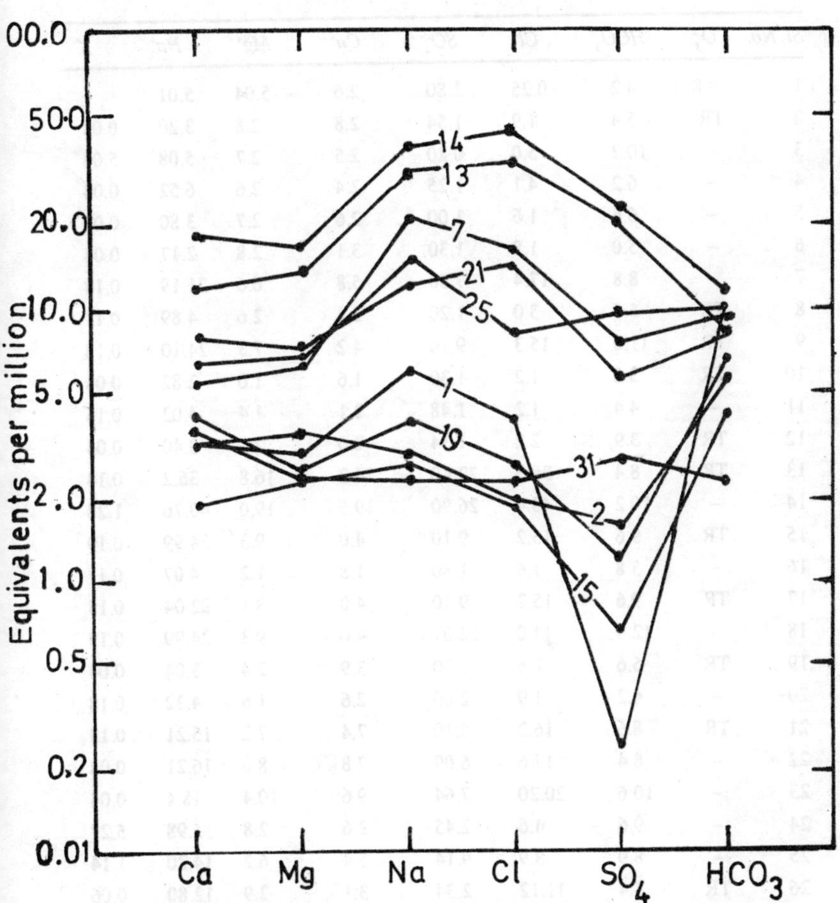

FIG.7.2 SCHOELLER NOMOGRAPH

Table 7.1 : Chemical Analysis of Groundwater from Coastal Sandstones (May, 1989)

Sl. No.	CO_3^-	HCO_3^-	Cl^-	SO_4^{++}	Ca^{++}	Mg^{++}	Na^+	K^+
1	–9.8	4.2	0.25	2.80	2.6	5.04	5.01	
2	TR	5.4	1.9	1.54	2.8	2.8	3.20	0.04
3	–	10.2	5.0	0.20	2.5	2.7	5.08	5.04
4	–	6.2	4.1	1.25	2.4	2.6	6.52	0.03
5	–	6.6	1.6	1.00	2.6	2.7	3.80	0.07
6	–	5.0	1.8	1.30	3.1	2.8	2.17	0.04
7	–	8.8	17.4	7.50	5.8	6.6	21.19	0.14
8	TR	5.6	3.0	1.20	2.2	2.6	4.89	0.11
9	TR	11.0	15.3	9.10	4.2	7.3	24.10	0.18
10	TR	3.0	1.2	1.30	1.6	1.0	2.82	0.04
11	–	4.4	1.2	1.48	2.1	1.4	4.02	0.16
12	TR	3.9	2.2	1.34	2.0	2.0	3.40	0.04
13	TR	8.4	36.5	22.40	14.2	16.8	36.2	0.14
14	–	10.2	43.4	26.90	19.5	19.0	40.76	1.28
15	TR	9.6	15.2	9.10	4.0	9.3	24.99	0.19
16	–	3.8	1.6	1.80	1.8	1.2	4.07	0.11
17	TR	9.6	15.2	9.10	4.0	8.1	22.04	0.15
18	–	12.4	14.0	12.01	4.0	9.3	24.99	0.19
19	TR	5.6	2.6	1.20	3.9	2.4	3.04	0.04
20	–	4.2	1.9	2.60	2.6	1.6	4.32	0.18
21	TR	8.6	16.2	5.10	7.4	7.2	15.21	0.13
22	–	8.4	17.6	6.09	7.8	8.0	16.21	0.08
23	–	10.6	20.20	7.64	9.6	10.4	18.4	0.04
24	–	9.6	4.6	2.45	2.6	2.8	5.98	5.27
25	–	8.9	8.9	9.14	5.4	6.2	14.20	1.14
26	TR	5.4	11.12	2.34	3.1	2.9	12.80	0.06
27	–	4.3	1.8	0.78	2.4	2.2	2.24	0.04
28	–	10.4	12.4	9.22	6.4	5.1	16.40	4.12
29	TR	4.1	2.2	2.00	1.8	2.1	4.32	0.08
30	TR	5.2	3.0	0.04	2.2	2.1	3.82	0.08
31	TR	2.3	2.6	3.40	3.1	2.4	2.17	0.08
32	–	3.2	1.2	1.20	1.9	1.0	11.41	0.11
33	–	3.9	2.2	0.73	1.9	2.1	2.83	0.03

Table 7.2 : Major ion ratios of groundwater from sandstones

Sl. No	Ca/ Mg	$\frac{Na+K}{Mg}$	$\frac{Na+K}{Ca}$	Cl/ Na+K	Cl/ Mg	Cl/ Ca	$Cl+SO_4$/ HCO_3	Symbol	Remarks
1	1.03	3.53	3.41	0.54	1.93	1.86	0.57	2331221	CDDE
2	0.97	0.90	0.93	1.00	0.90	0.93	0.83	1112111	CDDE
3	0.93	3.37	3.62	0.54	1.83	1.96	0.52	1331221	CDDE
4	1.03	2.08	2.01	0.72	1.50	1.44	0.76	2331221	CDDE
5	0.93	1.64	1.76	0.38	0.63	0.68	0.48	1221111	CDDE
6	1.06	0.65	0.61	1.13	0.74	0.69	0.65	2112111	CDDE
7	0.86	2.99	3.44	0.87	2.62	3.01	2.82	1331333	DDCS
8	0.89	1.81	2.04	0.79	1.44	1.62	0.83	1231221	DDCS
9	0.61	3.25	5.35	0.62	2.01	3.31	2.11	1331333	DDCP
10	1.50	2.13	1.42	0.74	3.58	1.05	0.91	2321321	CDDU
11	1.46	2.73	1.41	0.46	1.26	0.86	0.75	2321211	CDDU
12	1.09	1.01	1.93	1.03	1.04	0.96	0.95	2222211	CDDE
13	0.85	21.4	2.49	1.02	2.18	2.54	7.89	1332333	DDCP
14	1.02	2.20	2.16	1.04	2.29	2.24	6.70	2332333	CDDE
15	0.91	1.37	1.51	0.59	0.82	0.90	0.64	1221111	DDCS
16	1.35	2.83	2.08	0.48	1.35	0.00	0.95	2331211	CDDU
17	0.51	2.69	5.25	0.70	1.89	3.69	2.56	1331213	DDCP
18	0.42	2.62	6.17	0.58	1.53	3.60	2.16	1331233	CDDU
19	1.61	1.15	0.71	1.00	1.15	0.71	0.72	2212211	CDDU
20	1.59	2.55	1.60	0.46	1.17	0.74	0.82	2321211	CDDU
21	1.05	2.09	1.98	1.10	2.30	2.19	2.53	2322333	CDDE
22	0.88	2.70	2.02	1.12	2.21	2.26	2.82	1332333	DDCS
23	0.93	1.70	1.73	1.13	1.94	2.08	2.65	1222233	DDCS
24	0.86	3.70	4.29	0.45	1.65	1.92	0.75	1331221	DDCS
25	0.81	2.31	2.83	0.58	1.34	1.64	2.00	1331223	DDCS
26	1.10	4.14	3.76	0.95	3.93	33.56	2.51	2331333	CDDE
27	1.08	0.94	0.86	0.84	0.79	1.73	0.55	2111121	CDDE
28	1.29	3.96	3.06	0.62	2.45	1.89	2.04	2331323	CDDE
29	0.95	2.13	2.24	0.58	1.25	1.31	1.02	1331222	DDCS
30	0.96	1.59	1.66	0.86	1.37	1.43	0.62	1221221	DDCS
31	1.19	0.83	0.69	1.34	1.11	0.93	2.77	2112212	CDDE
32	1.67	9.40	5.61	0.12	1.17	0.70	3.44	2331213	CDDE
33	0.96	1.28	1.25	0.74	0.96	1.29	0.74	1221121	DDCS

CDDE : Calcite and Dolomite dissolution at equilibrium

DDCS : Dolomite Dissolution with Calcite Supersaturation

DDCP : Dolomite Dissolution with Calcite Precipitation

CDDU : Calcite Dissolution with probably Dolomite under saturation

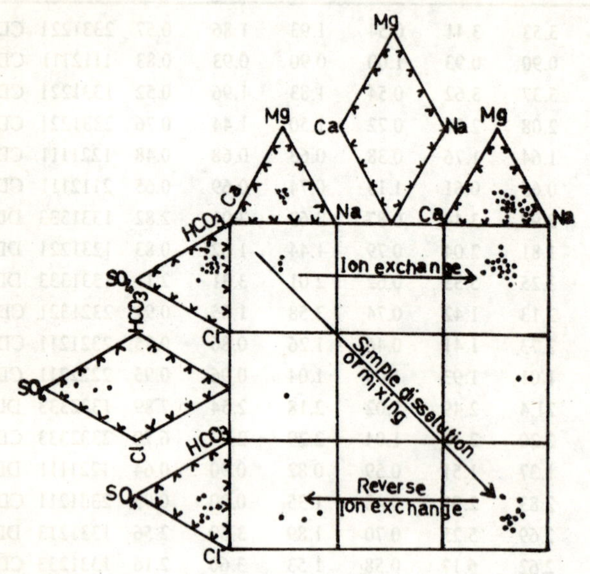

FIG.7-3 DUROV DIAGRAM WITH PROCESS AND WATER GROUPS

hydrochemical facies in the groundwater of the coastal sandstones are described here.

Field 1 : HCO_3^- and Ca^{2+} dominant, frequently indicates recharging waters in sandstones.

Field 3 : HCO_3^- and Na^+ dominant, normally indicates ionic exchanged waters.

Field 4 : SO_4^{2-} and Ca^{2+} dominant, here indicates a mixed water or simple solution.

Field 6 : SO_4^{2-} and Na^+ dominant, not frequently encountered, indicates a mixed water.

Field 7 : Cl^- and Ca^{2+} dominant, infrequently encountered, due to reverse ion exchange of Na^+ Cl^- waters.

Field 9 : Cl^- and Na^+ dominant, frequently indicates end point waters.

Spatial variation map thus prepared (Fig. 7.1) suggests the existence of sodium bicarbonate water in the southern, southeastern and southwestern parts of the sandstone belt. Northern part of the sandstone belt is dominated with sodium chloride water.

The chemical genesis of groundwater present in the area is also studied using modified Durov diagram, which is shown in Figure 7.3. From this figure, ion exchanged water is located in the southern, southeastern and southwestern parts and the old brackish water in the northern side of the sandstone belt. Fresh recharge water is confined to the local pockets. It supports the assumption that the old brackish waters are related to the paleo-saline water intrusion.

Since the clays are present adjacent to the sandstone belt on the northern side there is no possibility of fresh water being recharged into the sandstone belt. But the existence of fresh water recharge in the southern part of the sandstones is due to the characteristic fractures and joints present in this region. As a result, old brackish water is retained in the northern part of the sandstone belt, whereas it is diluted towards south due to fresh water recharge.

References

Burder, D.L., and Mazloum, S., (1958). Some chemical types of groundwaters from Syria. UNESCO Symp. Teheran, Paris, pp. 73-90.

Durov, S.A., (1948). Natural waters and graphic representation of their composition. Dokl. Akad. Nauk. SSR, V. 59, pp. 87-90.

Eischami, F. and Hamdan, A.R., (1989.) Hydrochemical classification of the groundwater of the Wadi-al Bihwell field in the northern part of the United Arab Emirates. International workshop on appropriate methodologies for development and management of groundwater resources in developing countries V. 2, pp. 853-65.

Haddad, R.H., (1981). A proposed numerical approach to the classification of groundwater. First symposium on the future development of water resources in the Gulf and Arabian Peninsula. Kuwait, Jour. Gulf Arab Penin. Stud. pp 47-61.

Hill, R.A., (1940). Geochemical patterns in the Coachella valley, California. Trans. American Geophy. Union, V. 21. pp. 46-49.

Lloyd, J.W., (1965). The hydrochemistry of the aquifers of north-eastern Jordan. J. Hydrol. V. 3, pp. 319-30.

Lloyd, J.W. and Heathcote, J.A., (1985). Natural inorganic hydrochemistry in relation to groundwater. An introduction. Clarendon Press, p. 296.

Piper, A.M., (1944). A graphic procedure in the geochemical interpretation of water analysis. Trans. Am. Geophys. Union. V. 25, pp. 914-23.

Ramteke, D.S. and Moghe, C.A., (1988). Manual on water and waste water analysis, NEERI, Nagpur, pp. 331.

Schoeller, H., (1959). Arid zone hydrology-Recent developments. Arid zone Research 12 UNESCO, Paris.

Srinivas, D., (1990). Sedimentology and hydrogeology of the coastal sandstone belt from Guntur district, Andhra Pradesh (M. Phil thesis submitted to Nagarjuna University). 87 pp.

8

Geochemical Studies of Hot Spring and Deep Bore Hole Waters of Manikaran, Himachal Pradesh, India

V.K. SAXENA

Introduction

Manikaran geothermal area is located in Parbati Valley, Himachal Pradesh and known as a famous geothermal field in India, where a large number of hot springs and bore holes with temperature ranging from 41 to 97°C occur. The thermal water samples were collected from six hot springs and three bore holes (120m) and studied for their aqueous ionic species that are useful for the study of mineral water equilibria and the reservoir temperature of the geothermal systems (Helgeson 1969, Kharaka and Barnes 1973, Norton and Panichi 1978, Arnorsson 1983, Giggenbach *et al.*, 1983 and Saxena and D'Amore 1984 and Saxena, 1987).

Manikaran (32°02'N 77°25'E) is located in North West Himalaya at an elevation of 1700 masl. The rock types exposed in this area are a sequence of thick massive quartzites with subordinate phyllites, belonging to the Precambrian meta sedimentary sequence. The heat anomaly exists within the jointed and fractured quartzites which are overlain by an overthrust of carbonaceous phyllites, limestones and mica schists, etc. The lithology of this area consists of travertine deposits, quartzites, granite gneisses and debris. Majority of the springs emerge on the bank of the Parbati River and Bramha Ganga River (Fig. 8.1).

Material and Methods

The Chemical analyses of the water samples were performed by different techniques, as follows :

Ca, Mg : EDTA and Atomic Absorption Spectometry.

Na, K, Li : Atomic Absorption Spectometry.

HCO_3 : Acidimetric.

SO_4 : Gravimetric.

Cl : Mohr and Chloride meter.

B : Distillation followed by Mannito-Boric titrametric.

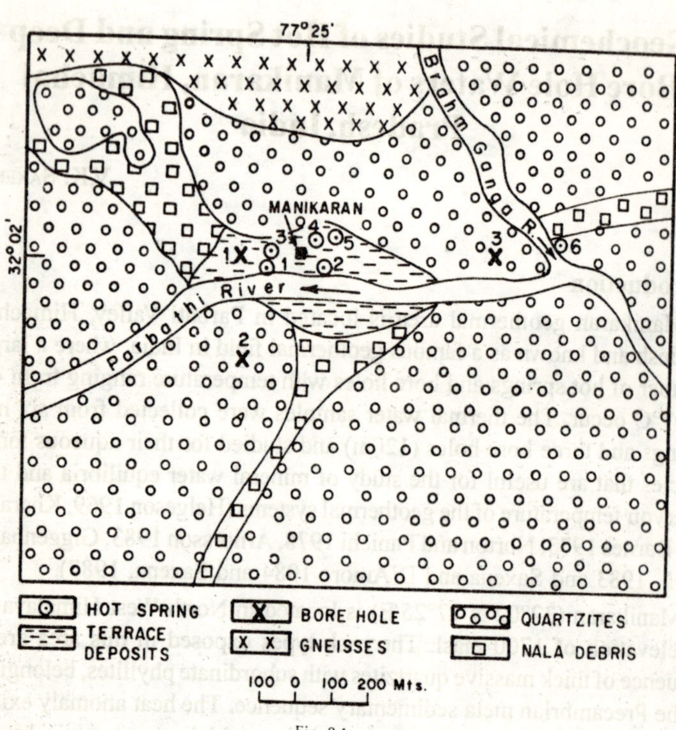
77°25'
32°02'
MANIKARAN
Brahm Ganga R.
Parbathi River
HOT SPRING
TERRACE DEPOSITS
BORE HOLE
GNEISSES
QUARTZITES
NALA DEBRIS
100 0 100 200 Mts.

Fig. 8.1

Table 8.1 : Chemical analyses of the Manikaran geothermal waters (concentration in mg/lit.)

Area	*Temp.* (°C)	*Spe. Cond.*	*pH*	*TDS*	*SiO_2*	*Na*	*K*	*Ca*	*Mg*	*Cl*	*RCO_3*	*SO_4*	*Li*	*B*
MH–1	96	974	7.4	672	118	102	21	65	11.5	97	234	42	1.2	1.8
MH-2	97	878	7.3	555	120	94	22	42	10.0	123	177	45	1.2	2.2
MH-3	87	931	7.5	633	115	92	19	46	4.5	76	240	34	1.0	2.0
MH-4	81	864	7.4	596	120	88	18	42	11.0	104	175	45	0.8	2.0
MH-5	45	780	7.5	530	110	75	15	42	10.0	64	178	46	0.6	1.0
MH-6	77	752	7.5	535	105	66	15	45	4.5	65	180	41	0.5	2.0
MB-1	77	647	7.0	446	100	52	10	46	8.0	59	147	32	0.2	0.5
MB-2	95	730	7.8	503	108	58	10	45	20.0	83	148	48	0.2	0.7
MB-3	70	720	7.5	494	98	50	9	45	22.0	50	169	40	0.2	0.7

MH : Manikaran Hot spring. MB : Manikaran bore hole.

surface temperature.

specific conductivity in micromhos/cm at 25°C,

pH at 25°C.

TDS : Total Dissolved Solids

SiO_2 : Atomic Absorption and Spectrometric (Molibdate blue)
TDS : Sum of total cations and anions.

Analyses were performed by Atomic Absorption Spectometry, using Varian Techtron. Spectrophotometric determinations were carried out by Systronics-102 and Spectronic-20.

Results and Discussion

The S^2H and $S^{18}O$ compositions of geothermal waters discharged from hot springs and bore-holes are shown in Fig. 8.2. This figure shows that MH-4, MH-5, MH-6, MB-1 and MB-2 are close to fit a meteoric water line of $S^2H = S^{18}O + 9.5$ which is more representative of meteoric water. The variation of S^2H and $S^{18}O$ in MH-3, MH-1 and MB-3 may suggest an increase in heavy isotopes during underground meteoric water percolation.

Chemical analyses of waters from the bore holes and springs (Table 8.1) indicate that :

1. These thermal waters are nearly neutral to mildly alkaline in pH (7.3 to 7.8) and classified as Na-Ca-HCO_3-Cl type.

2. Hot spring waters contain comparatively more total dissolved solids than bore holes.

3. Chloride is more in hot spring than in bore holes.

4. Li and B contents are comparatively less in bore holes.

The above characteristics and the chemical composition of Manikaran

Table 8.2 : Equilibrium indices of Anhydrite, Calcite, Dolomite, Fluorite, Magnesite and Quartz.

Area	*Anhydrite*	*Calcite*	*Dolomite*	*Fluorite*	*Magnesite*	*Quartz*
MH-1	-1.513	0.7844	0.5565	-3.4967	-0.6656	0.3835
MH-2	-1.4927	0.4542	-0.0783	-1.1748	-0.9683	0.3850
MH-3	-1.6287	0.7531	1.0913	-5.4232	-0.1172	0.4725
MH-4	-1.7084	0.2037	0.5708	-1.5008	-0.1004	0.5643
MH-5	-2.0287	0.0229	0.1586	-3.4261	-0.4162	0.9596
MH-6	-1.5857	0.5396	0.6824	-0.8496	-0.3329	0.5483
MB-1	-1.7194	0.6849	1.2939	-5.7133	0.1333	0.5156
MB-2	-1.6122	0.6632	1.6228	-1.7595	0.4795	0.5728
MB-3	-1.7201	0.3390	1.0574	-6.8793	0.2269	0.6004

Calculated by using Kharaka and Barnes 1973, "Solmneq" computer programme.

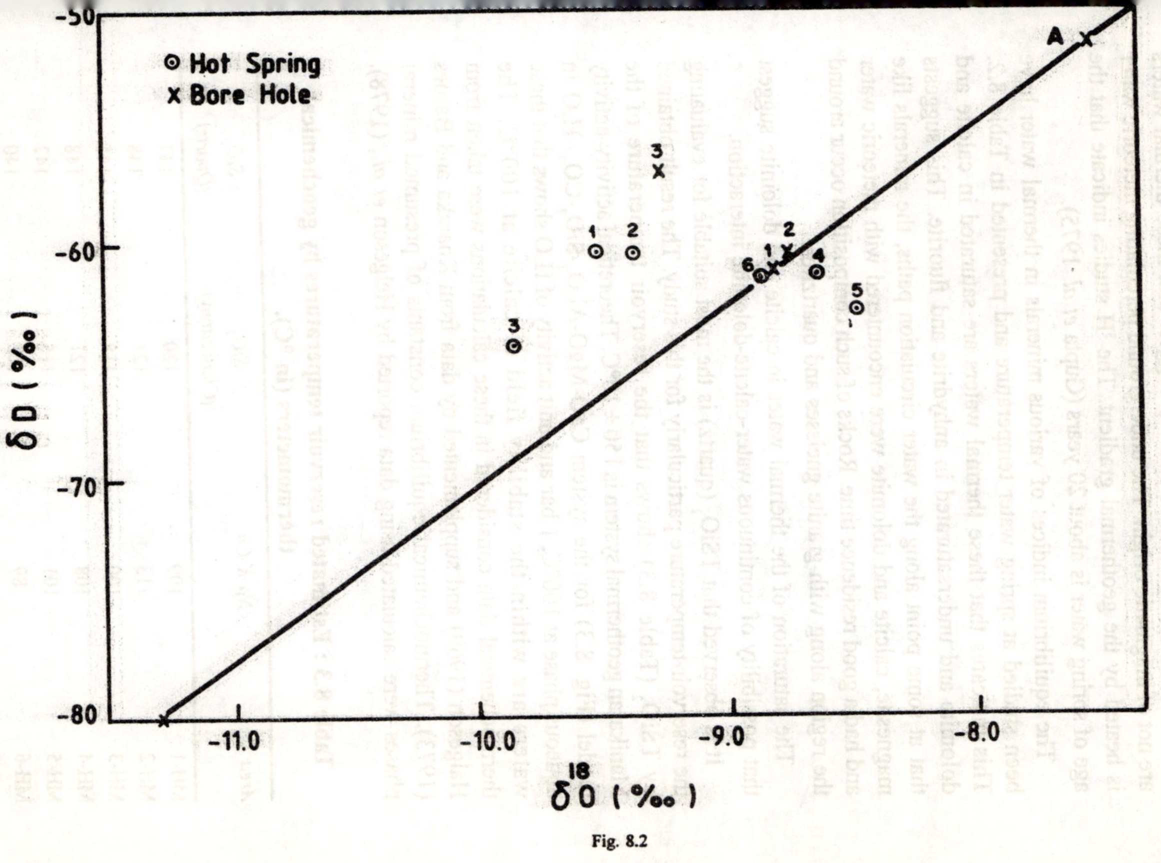

Fig. 8.2

hot spring and bore hole waters may suggest that these thermal waters are not of magmatic origin. The underground percolating meteoric water is heated by the geothermal gradient. The ^3H studies indicate that the age of spring water is about 20 years (Gupta *et al.* 1975).

The equilibrium indices of various minerals in thermal water have been studied at spring water temperature and presented in Table 8.2. This indicates that these thermal waters are saturated in calcite and dolomite and undersaturated in anhydrite and fluorite. This suggests that at some point along the water circulation paths, the minerals like magnesite, calcite and dolomite were encountered with meteoric water and had a good residence time. Rocks of such composition occur around the region along with granite gneisses and quartzites.

The saturation of the thermal water in calcite and dolomite suggest that possibility of continuous water-calcite-dolomite interaction.

It is observed that $TSiO_2$ (quartz) is the most suitable for evaluating the reservoir temperature particularly for this study. The result obtained by $TSiO_2$ (Table 8.3) shows that the reservoir temperature of the Manikaran geothermal system is 150 ± 10°C. Theoretical activity-activity model (Fig. 8.3) for the system CaO-MgO-Al_2O_3-SiO_2-CO_2. H_2O in aqueous phase at 100°C, 1 bar and unit activity of H_2O shows that these waters are within the stability field of calcite at 100°C. The thermochemical data considered in these calculations were taken from Halgeson (1969) and supplemented by data from Kharaka and Barnes (1973). Thermochemical equilibrium constants of presumed mineral phases were calculated using data reported by Helgeson *et al.*, (1978).

Table 8.3 : Estimated reservoir temperatures by geochemical thermometers (in °C).

Area	*Na-K-Ca*	*SiO_2 (Chalcedony)*	*SiO_2 (Quartz)*
MH-1	109	120	147
MH-2	113	121	148
MH-3	100	118	145
MH-4	108	121	148
MH-5	101	115	142
MH-6	89	112	140
MB-1	75	109	137
MB-2	76	114	142
MB-3	73	108	136

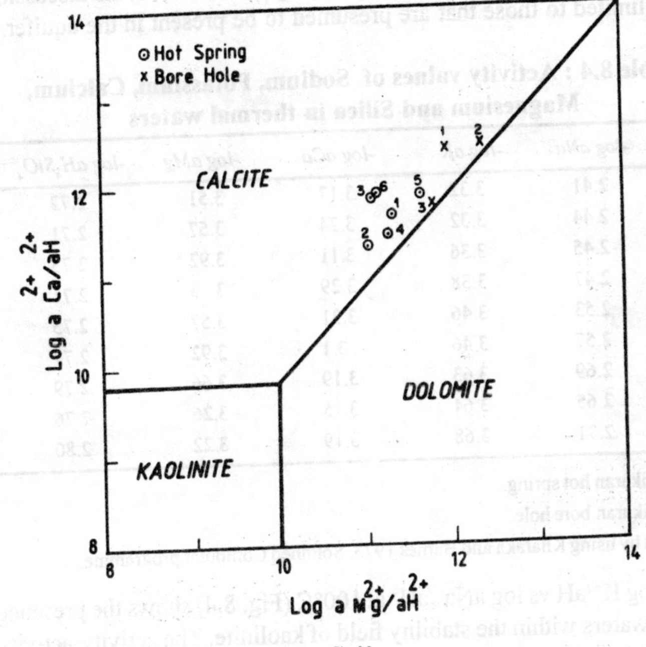
⊙ Hot Spring
x Bore Hole
CALCITE
DOLOMITE
KAOLINITE
Log a Ca2+/aH2+
Log a Mg2+/aH2+
14
12
10
8
8
10
12
14
1
2
3
4
5
6
1
2
3

Fig. 8.3

The resultant (computer calculated) activity values are presented in Table 8.4. As the number of minerals which could participate in a thermochemical reaction with the aqueous phase are more, the discussion here is limited to those that are presumed to be present in the aquifer.

Table 8.4 : Activity values of Sodium, Potassium, Calcium, Magnesium and Silica in thermal waters

Area	*-log aNa*	*-log aK*	*-log aCa*	*-log aMg*	*-log aH_4SiO_4*
MH-1	2.41	3.32	3.17	3.51	2.72
MH-2	2.44	3.32	3.24	3.57	2.71
MH-3	2.45	3.36	3.11	3.92	2.73
MH-4	2.47	3.38	3.29	3.53	2.71
MH-5	2.53	3.46	3.31	3.57	2.75
MH-6	2.57	3.46	3.1	3.92	2.77
MB-1	2.69	3.63	3.19	3.66	2.79
MB-2	2.65	3.64	3.15	3.26	2.76
MB-3	2.71	3.68	3.19	3.22	2.80

MH : Manikaran hot spring.

MB : Manikaran bore hole.

Calculated by using Kharaka and Barnes 1973, Solmneq Computer programme.

The log K^+/aH vs log aNa^+/aH^+ at 100°C (Fig. 8.4) shows the presence of those waters within the stability field of kaolinite. The activity-activity model for log aNa^+/aH^+ vs log aH_4/SIO_4 (Fig. 8.5) shows that the thermal waters are apparently in equilibrium with quartz and within the stability field of Na-Montmorillonite at 150°C . The log aK/aH vs log $aHSO_4$ (Fig. 8.6) shows that these waters are in equilibrium with kaolinite, montmorillonite and quartz at 150°C.

From these models, it can be concluded that these thermal waters are apparently equilibrated at 150°C with quartz, montmorillonite and kaolinite.

Conclusions

1. The Manikaran thermal waters are of meteoric water origin and of Na-Ca-HCO_3-Cl type.
2. The geothermal waters are heated by normal geothermal gradient and deep circulation.
3. $TSiO_2$ and chemical equilibria gives 150±10°C as reservoir temperature.

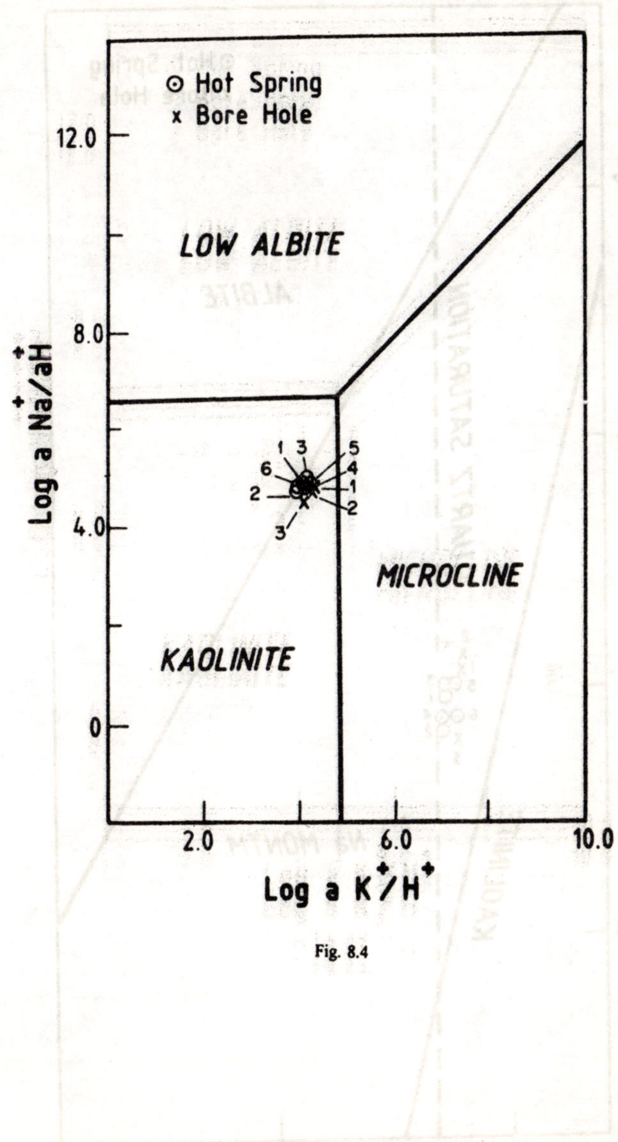
⊙ Hot Spring
x Bore Hole
LOW ALBITE
MICROCLINE
KAOLINITE
12.0
8.0
4.0
0
2.0
6.0
10.0
Log a Na⁺/aH⁺
Log a K⁺/H⁺
1
3
5
6
4
2
1
2
3

Fig. 8.4

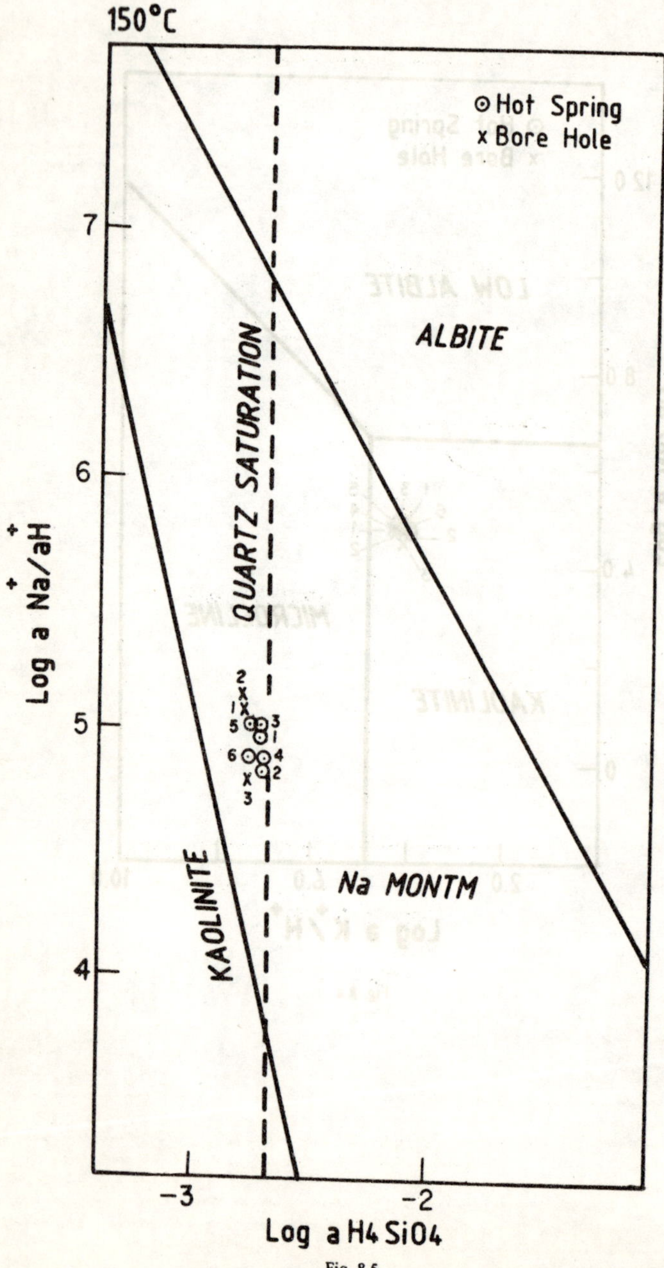
150°C
⊙ Hot Spring
x Bore Hole
ALBITE
QUARTZ SATURATION
KAOLINITE
Na MONTM
Log a Na+/aH+
Log a H4 SiO4
7
6
5
4
-3
-2

Fig. 8.5

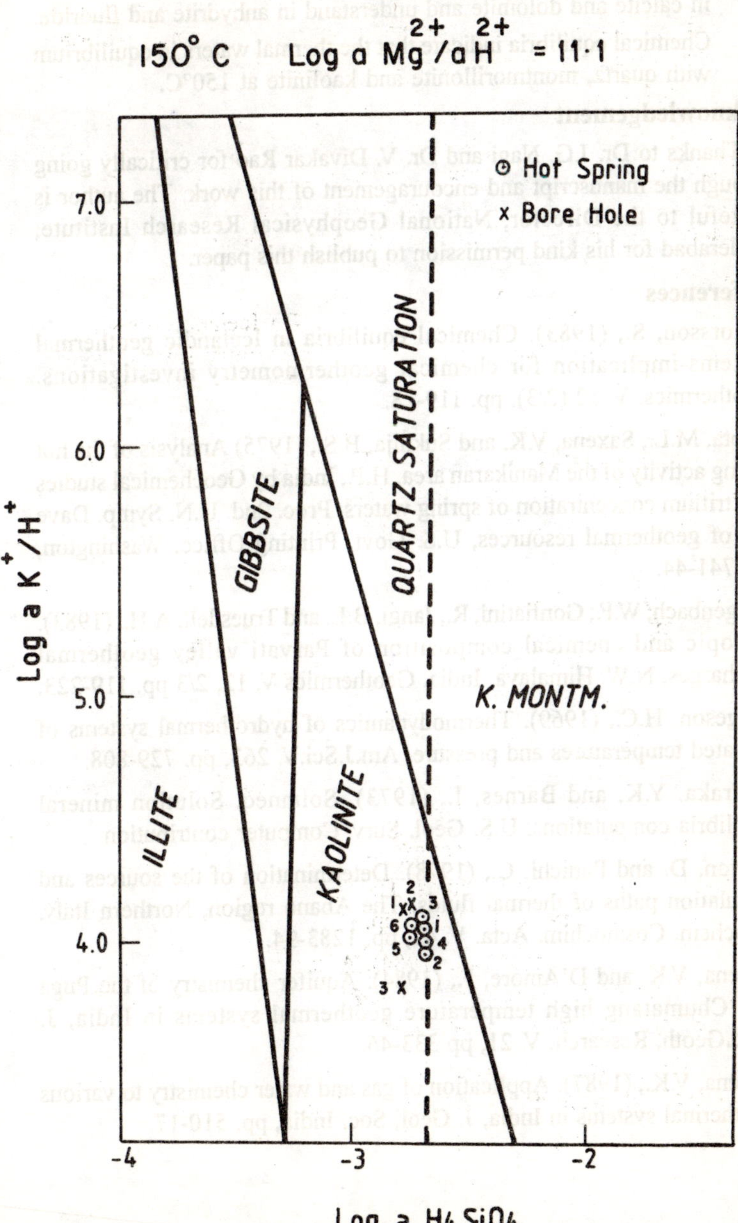
150° C Log a Mg2+/aH2+ = 11·1
⊙ Hot Spring
x Bore Hole
7.0
6.0
5.0
4.0
Log a K+/H+
QUARTZ SATURATION
GIBBSITE
K. MONTM.
KAOLINITE
ILLITE
-4
-3
-2
Log a H4 SiO4

Fig. 8.6

4. Equilibrium indices shows that these hot spring waters are saturated in calcite and dolomite and understand in anhydrite and fluoride.
5. Chemical equilibria indicate that the thermal waters in equilibrium with quartz, montmorillonite and kaolinite at 150°C.

Acknowledgement

Thanks to Dr. J.G. Nagi and Dr. V. Divakar Rao for critically going through the manuscript and encouragement of this work. The author is grateful to the Director, National Geophysical Research Institute, Hyderabad for his kind permission to publish this paper.

References

Arnorsson, S., (1983). Chemical equilibria in Icelandic geothermal systems-implication for chemical geothermometry investigations. Geothermics, V. 12 (2/3), pp. 119-28.

Gupta, M.L., Saxena, V.K. and Sukhija, B.S., (1975) Analysis of the hot spring activity of the Manikaran area, H.P., India by Geochemical studies and tritium concentration of spring waters. Proc. 2nd. U.N. Symp. Dave use of geothermal resources, U.S. Govt. Printing Office, Washington, pp. 741-44.

Giggenbach, W.F., Gonfiatini, R., Jangi, B.L. and Truesdell, A.H., (1983). Isotopic and chemical composition of Parvati valley geothermal discharges, N.W. Himalaya, India. Geothermics V, 12, 2/3 pp. 119-223.

Helgeson, H.C., (1969). Thermodynamics of hydrothermal systems of elevated temperatures and pressure. Am.J.Sci.V. 263, pp. 729-808.

Kharaka, Y.K. and Barnes, I., (1973). Solmneq. Solution mineral equilibria computations, U.S. Geol. Surv. Computer contribution.

Norton, D. and Panichi, C., (1978). Determination of the sources and circulation paths of thermal fluids. The Abano region, Northern Italy, Geochem. Cosmochim. Acta. V. 42, pp. 1283-94.

Saxena, V.K. and D'Amore, F., (1984). Aquifer chemistry of the Puga and Chumatang high temperature geothermal systems in India, J. Volc.Geoth, Research. V. 21, pp 333-46.

Saxena, V.K., (1987). Application of gas and water chemistry to various geothermal systems in India, J. Geol, Soc. India, pp. 510-17.

9

Groundwater Quality from Guntur Municipal Corporation, Andhra Pradesh

K. Rajendra Rao
P. Sankara Pitchaiah

Introduction

The term "Water Quality" is an extremely confusing concept, meaning different things to different people. In groundwater resource evaluation, the quality of groundwater is of nearly equal importance to quantity. A knowledge of water quality is also essential for rational management of water and land resources (Karanth 1989). Here, mainly chemical parameters are studied to assess the water quality.

Methodology

Forty-nine groundwater samples are collected systematically from various parts of the Guntur Municipal Corporation (GMC) area and its surroundings (Fig. 9.1, Sankara Pitchaiah and Rajendra Rao, 1998). These samples are subjected for chemical analysis and the results are presented in Table 9.1.

Chemical Parameters

pH

The studied samples of water are found to have pH values ranging between 7.1 and 8.1, that are in the alkaline range. The minimum is found in R. Agraharam, and Koretapadu wells while the maximum in Syamalanagar well.

In general, groundwater has pH values between 5 and 8. Very high pH values, above 8.5, are usually associated with sodium carbonate and bicarbonate water, moderately high pH waters with bicarbonates and very low pH values may be associated with small amounts of mineral acids from sulphide sources or with organic acids. Most of the samples collected from GMC fall in the medium range.

Electrical Conductivity (EC)

The EC values ranging between 340 micro-mhos/cm and 22,100 micro mhos/cm. The minimum value is recorded in Lalapet well, and the maximum in Autonagar well. Most of the wells have high EC value and indicates saline nature.

Table 9.1 : Chemical Parameters in Groundwater

Sl. No.	Location	pH	Conductance micro-mohs/cm	T.D.S. mg/l	Alkalinity mg/l	Hardness mg/l	Carbonate mg/l	Bicarbonate mg/l	Nitrate mg/l	Chloride mg/l	Sulphate mg/l	Calcium mg/l	Magnesium mg/l	Sodium mg/l	Potassium mg/l
1.	Krishna Nagar	8.4	1397	894.08	384	415	–	468.48	15.06	251.92	168.7	24	60.51	176	11
2.	Lakshmipuram	9	1434	917.76	712	160	28.8	810.08	42.52	77.98	161.25	5.33	47.82	321	6
3.	Ankireddy Palem	8.8	4380	2803.2	508	330	24	570.96	170.11	729.77	537.5	24	196	950	7
4.	Srinivasaraothota	8.3	2940	1881.6	336	580	–	409.92	4.78	673.79	276.25	26.67	104.76	480	9
5.	R. Agraharam	7.9	3960	2534.4	340	900	–	340	366.8	755.77	410	40	14.08	530	66
6.	Ravindranagar	7.8	2740	1753.6	248	1045	–	248	199.35	565.82	295	117.33	99.23	200	10
7.	By-Pass Road	7.8	13670	8748.8	396	1885	–	396	124.48	91.35	VH	106.67	276.53	1960	8
8.	Pattabhipuram	7.7	2850	1824	276	780	–	276	17.46	603.81	312	37.33	144.77	233	12
9.	A.T. Agraharam	7.9	9470	6060.8	724	895	24	724	545.78	1159.6	228	21.33	169.17	2040	8
10.	Koretapadu	7.8	3250	2080	244	1336	–	244	279.98	767.76	VH	93.33	195.2	166	14
11.	Kannavarithota	7.5	1787	1143.7	304	645	–	304	164.8	3478.9	268	18.67	86.21	237	9
12.	Rajendra Nagar	7.6	907	580.48	232	415	–	232	44.74	121.96	207	34.67	53.68	82	4
13.	S.V.N. Colony	8.2	1462	935.68	480	610	–	440	159.48	183.94	154.5	18.67	68.32	214	7
14.	Choudary Pet	8.1	725	464	176	260	–	176	26.14	139.96	157.8	26.67	47.17	42	4
15.	Barathpet	7.8	3420	2188.8	680	610	–	680	202.01	529.84	421.9	13.33	81.33	710	6
16.	Mirch Yard	7.7	15510	9926.4	244	2325	–	244	166.57	3276.9	VH	370.67	461.97	1960	7
17.	Vidyanagar	8.1	2200	1408	324	520	–	324	60.69	459.86	1831.1	21.33	99.23	237	6.15
18.	L.I.C. Colony	7.9	1050	672	312	365	–	312	56.26	159.95	15.3	72	50.43	87	11
19.	Nagarampalem	7.7	1485	950.4	344	545	–	344	130.24	199.94	232	21.33	65.07	184	15
20.	Syamalanagar	8	1501	960.64	288	440	–	288	67.34	251.92	187.1	21.33	78.08	133	5
21.	Arundelpet 17 Line	8.1	1053	673.92	388	235	–	388	47.84	89.87	169.6	13.33	40.67	163	4
22.	Venugopal Nagar	8	2130	1363.2	500	350	–	500	37.21	321.9	248.75	13.33	30.91	460	3
23.	Gujjanagundla	7.7	5490	3513.6	444	625	–	444	547.55	951.7	476	26.67	91.09	590	1.82
24.	Kakani Road	7.5	1283	821.12	380	353	–	380	30.57	195.94	195	88	32.53	128	29
25.	Sampath Nagar	7.6	9060	5798	360	1835	–	360	76.19	2469.2	VH	218.67	258.64	1450	10

26. Prathapagiri Street	7.8	1920	1228.8	448	1340	–	448	163.02	275.91	242.5	34.67	76.45	176	112
27. Arundal Pet 2/5	7.7	2830	1811.2	380	970	–	380	317.18	445.86	346.25	21.33	185.33	283	20
28. Rajagari Thota	7.6	1479	946.55	592	290	–	592	36.33	145.95	197	8	60.19	245	11
29. B.R. Stadium	7.6	1600	1024	360	350	–	360	68.22	285.11	181	21.33	65.07	180	13
30. Hindu College	7.9	1540	985.6	440	320	–	440	72.56	197.94	197.7	16	52.05	223	12
31. Brodipet 2/4	7.9	761	487.04	288	255	–	288	29.24	79.98	167	32	30.91	67	7
32. M.D. Johny House	7.5	2490	1593.6	532	380	–	532	255.17	319.9	288.25	53.33	39.14	286	17
33. Sangadi Gunta	7.6	2960	1894.4	460	795	–	460	441.23	463.53	268	53.33	122	305	27
34. Mangaladas Nagar	7.7	1494	956.16	352	845	–	352	14.62	229.93	285	40.67	222	11	
35. Lalapet	7.6	2980	1907.2	676	425	–	676	333.14	306.57	442.9	16	82.96	630	16
36. Autonagar	7.4	26800	17152	376	4875	–	376	23.48	8740.6	VH	184	907.68	3520	16
37. Venkata Rao Pet	7.4	5670	3628.8	681	740	–	684	517.42	1182.9	VH	66.67	131.76	1120	19
38. Sangadi Gunta	7.3	2620	1676.8	520	650	–	520	214.41	373.22	326.2	56	122	291	66
39. Buchiah Thota	7.5	1582	1012.5	500	500	–	500	88.16	189.94	240.7	32	35.78	200	9
40. Rail Pet	7.8	2550	1632	480	805	–	480	379.21	306.57	291	37.33	123.62	237	14
41. Polichetty Tobaco's	7.8	2150	1376	224	465	–	224	45.19	289.91	522.8	56	69.95	237	13
42. Old Guntur	7.2	4640	2969.6	344	720	–	344	478.44	949.71	375.5	48	125.25	610	74
43. Sambasiva Pet 4th Line	7.4	1441	922.24	324	360	–	324	86.83	206.6	185	40	42.29	156	8
44. Nehru Nagar	7.7	1440	921.6	344	330	–	344	42.08	183.28	249.5	18.67	55.31	200	2
45. Old Guntur	7.4	3520	2252.8	264	995	–	264	303.01	666.46	482	109.33	125.25	286	65
46. Pandari Puram 6th Line	7.8	1979	1266.6	580	270	–	580	49.17	206.6	216.9	24	37.41	344	4
47. Kotha Pet	7.5	2560	1638	428	575	–	428	516.2	336.56	308	48	87.83	286	16
48. Chandra Mouli Nagar	7.5	1550	974.6	472	505	–	445	433.5	285.91	242.5	34	102.5	155	7
49. Kankara Gunta	7.2	3125	1824	410	710	–	460	543.4	482	VH	36	151.2	262	16

VH : Very High

Total Dissolved Solids (TDS)

Concentration of dissolved solids is an important parameter in drinking water and other quality standards. They give a particular taste, to the water, at higher concentration, also reduce its palatability.

TDS values range from 218 to 14,014 mg/l. Minimum value is found in Lalapet and maximum in Autonagar. Fifty per cent of samples have TDS values above the permissible limit (1,500 mg/l). (Fig. 9.1)

Alkalinity

Alkalinity value ranges from 95 mg/l to 930 mg/l. The minimum value is found in Syamalanagar well water and the maximum in A.T. Agraharam.

Hardness

Hardness of the water is property attributable to the presence of alkaline earths (Brown *et al.*, 1970). The major alkaline earth elements present in most natural waters are Ca and Mg. Other elements such as Sr/Ba (also alkaline earths), Fe, Mn and Al are normally not present in waters in sufficient amount to affect a test for hardness.

The recorded minimum and maximum hardness values in the study area are 12.5 and 87.5 mg/l respectively. Highest value is recorded from Autonagar well and the lowest from Lalapet.

From the hardness classification (Sawyer, 1960), it is understood that medium hard groundwater is distributed in Syamalanagar and Lalapet (Fig. 9.2). Hard water is located in Chowdarypet, Arundelpet, Rajagari thota, Brodipet, M.D. Johny house, Sambasivapet, Nehru nagar and Pandari puram areas. Very hard water is located in the remaining parts of the area.

Carbonates and Bicarbonates

Carbonates are generally present at very low levels, ranging from 30 to 170 ppm. Bicarbonates are found to range between 5 ppm (Syamalanagar) to 780 ppm (A.T. Agarjaram).

Chloride

Chloride ion is generally present in natural waters. The minimum and maximum chloride values are 24 mg/l and 7,381 mg/l respectively. Highest value is recorded in Autonagar well and the lowest in Lalapet. Fifty per cent of the wells have chloride concentration more than 250 mg/l. Very high concentration (>1,000 mg/l) is present in Bypass Road, A.T. Agraharam, Mirchiyard, Autonagar, Venkata Rao Pet areas (Fig. 9.3).

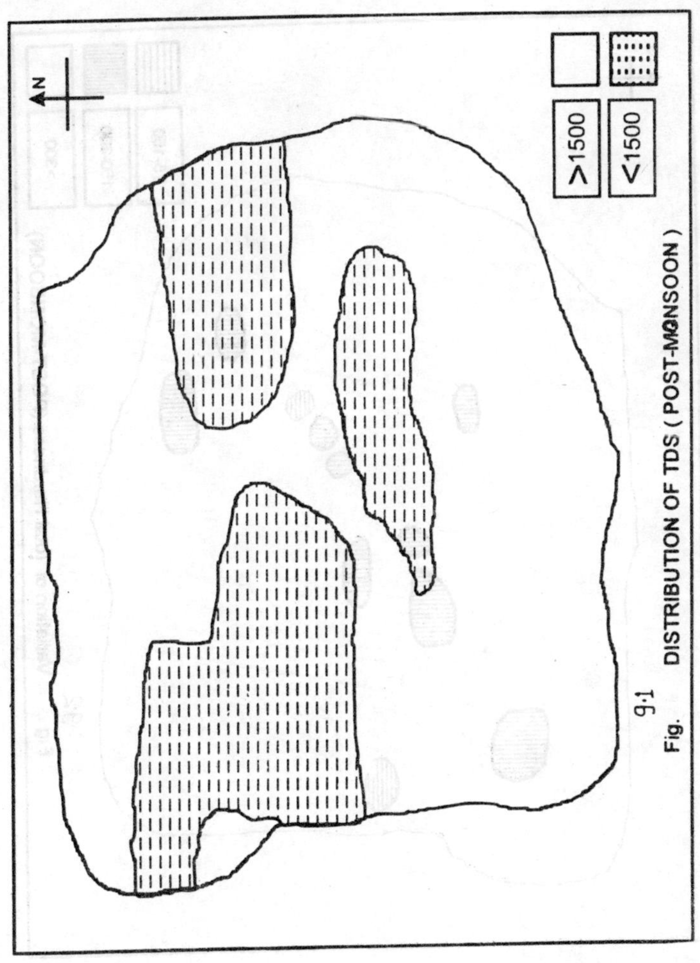

Fig. 9.1 DISTRIBUTION OF TDS (POST-MONSOON)

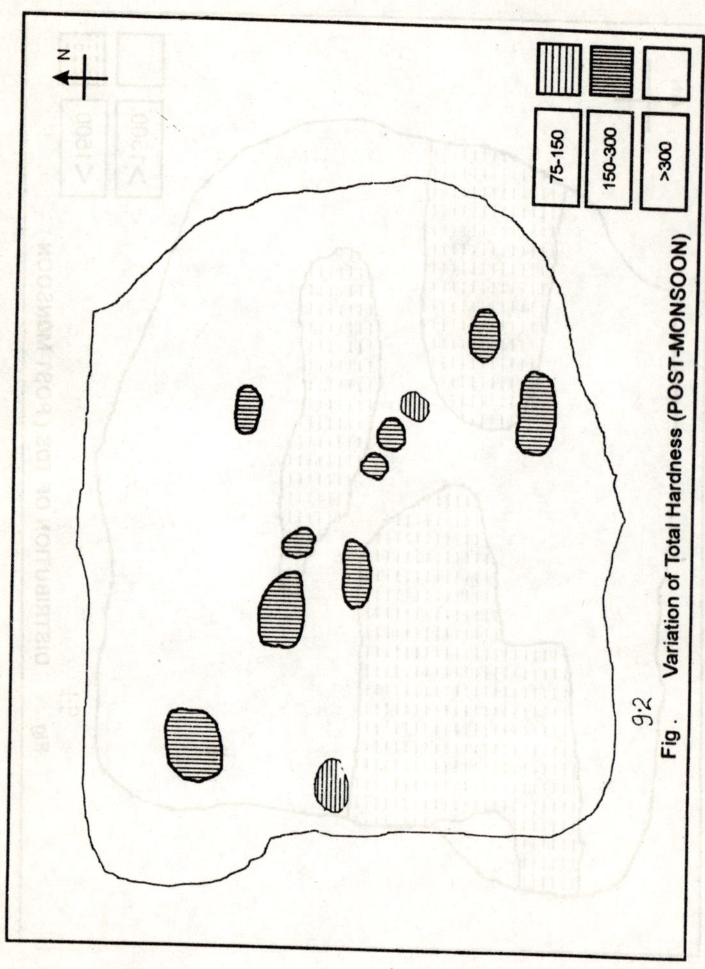

Fig. 9.2 Variation of Total Hardness (POST-MONSOON)

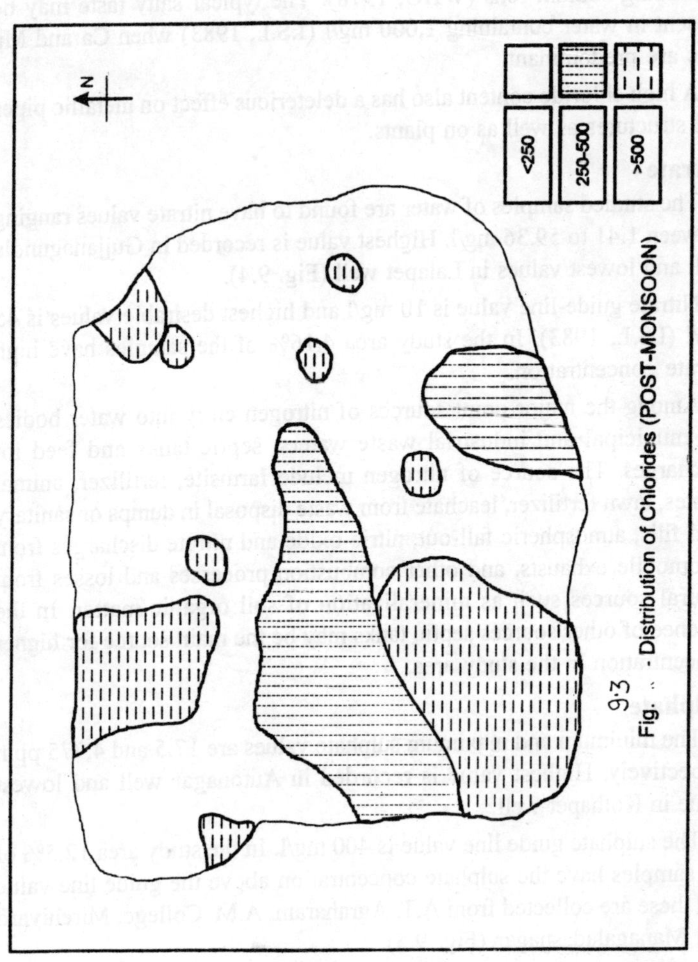

Fig. 9.3 Distribution of Chlorides (POST-MONSOON)

A concentration of 250 mg/l may be detectable in some waters containing sodium ions (WHO, 1978). The typical salty taste may be present in water containing 1,000 mg/l (I.S.I., 1983) when Ca and Mg ions are predominant.

A high chloride content also has a deleterious effect on metallic pipes and structures as well as on plants.

Nitrate

The studied samples of water are found to have nitrate values ranging between 1.41 to 59.36 mg/l. Highest value is recorded in Gujjanagundla well and lowest values in Lalapet well (Fig. 9.4).

Nitrate guide-line value is 10 mg/l and highest desirable values is 45 mg/l (I.S.I., 1983). In the study area 4.16% of the samples have high nitrate concentration.

Among the major point sources of nitrogen entry into water bodies are municipal and industrial waste waters, septic tanks and feed lot discharges. The source of nitrogen include farmsite, fertilizer, animal wastes, lawn fertilizer, leachate from waste disposal in dumps or sanitary land fills, atmospheric fall-out, nitric-oxide and nitrate discharges from automobile exhausts, and other combustion processes and losses from natural sources such as mineralization of soil organic matter. In the absence of other sources septic tanks may be the main source for higher concentration in the study area.

Sulphate

The minimum and maximum sulphate values are 17.5 and 4,775 ppm respectively. Highest value is recorded in Autonagar well and lowest value in Kothapet well.

The sulphate guide line value is 400 mg/l. In the study area 12.5% of the samples have the sulphate concentration above the guide line value and these are collected from A.T. Agraharam, A.M. College, Mirchiyard and Managaladasnagar (Fig. 9.5).

Calcium

The minimum and maximum calcium values are 10 and 460 mg/l respectively. Highest value is recorded in Mirchiyard well and lowest in Srinivasanagar colony area.

Magnesium

The minimum and maximum magnesium values in the study area are 7.32 mg/l and 1,220 mg/l respectively. The lowest value is recorded in Rajagari thota and highest value in Mirchiyard.

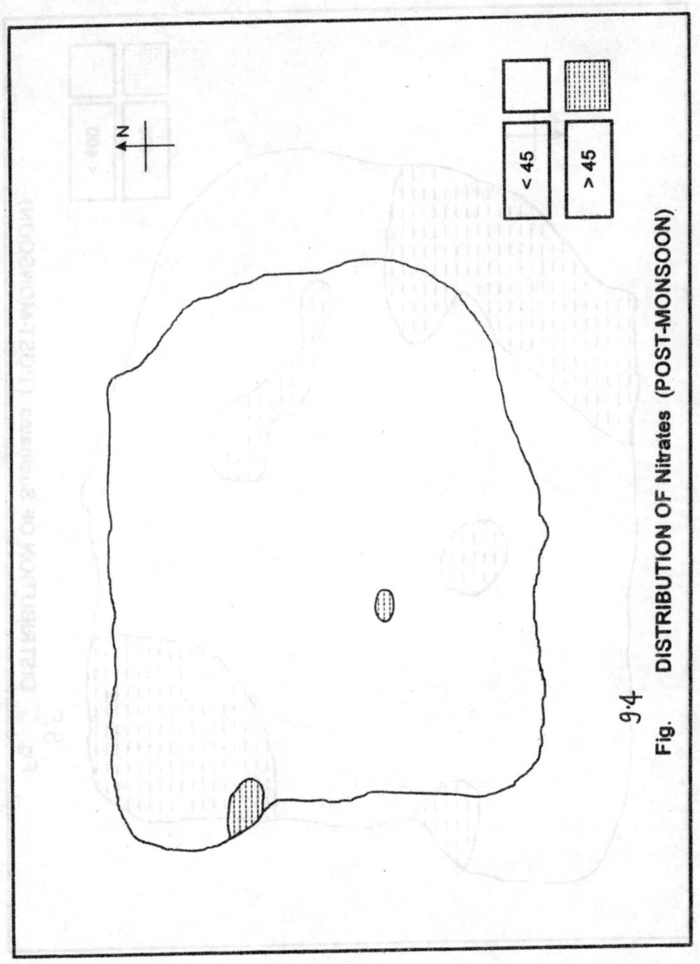

Fig. 9.4 DISTRIBUTION OF Nitrates (POST-MONSOON)

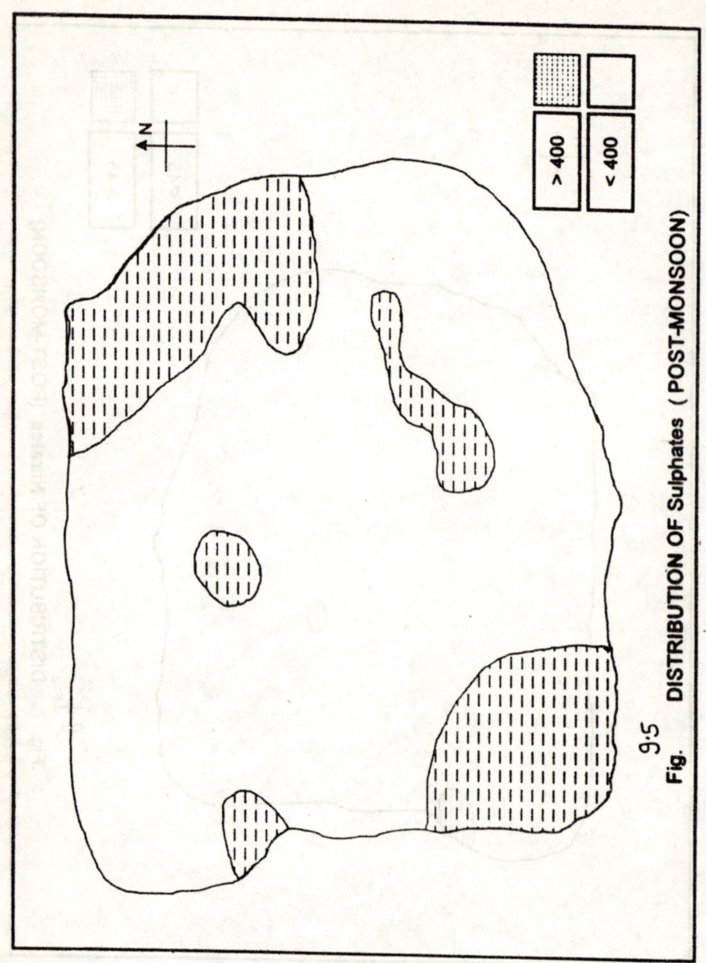

Fig. 9.5 DISTRIBUTION OF Sulphates (POST-MONSOON)

Water Quality for Irrigation

Though Guntur is an urban area, it is surrounded by number of villages like Takkelapadu, Uppalapadu, Perecherla, Ankireddypalem, Budampadu, Gorantla, Jonnalagadda, Kornepadu. In the outskirts of the town, crops like cotton, mirchi, paddy, jowar and tobacco are sown. In the town area groundwater is extensively utilizing for lawns, fruit and flower gardens. Hence, suitability of groundwater for irrigation is also discussed in this study.

Salinity

The minimum and maximum electrical conductivity values in the collected samples are 340 and 22,100 micro mhos/cm. According to the U.S. salinity laboratory (1954) classification there is no low salinity water in the study area (Fig. 9.6). Ten per cent of the samples collected belong to medium salinity water. It can be used, if a moderate amount of leaching occurs. Plants with moderate salt tolerance can be grown in most instances without special practices for salinity control. Thirty-nine per cent of samples collected are highly saline. This water cannot be used on soils with restricted drainage. Special management for salinity control may be required and plants with salt tolerance should be selected. The remaining samples fall under very high salinity class. This water is not suitable for irrigation under ordinary conditions, however, it may be used occasionally under very special circumstances, such as permeable soil, adequate drainage, application of excess irrigation water. Salt tolerant crops should be selected, while using this water.

Sodicity

It is assessed using SAR. Soil properties deteorated with increasing SAR levels in irrigation water (Sree Ram Murthy, 1989). It is found that the values vary in between 1.414 and 30.347. Eight per cent of water samples are good for irrigation. Four per cent of water samples are medium for irrigation. Others are bad for irrigation.

Carbonate and Bicarbonate ions

To quantify the effects of carbonates and bicarbonates Eaton (1950) proposed Residual sodium carbonate (R.S.C.) The minimum and maximum values from the water samples are –115.75 and 10.87 meq/l. According to the Eaton's classification seventy-six per cent of water samples are suitable for irrigation. Two per cent of samples collected from S.V.N. Colony have the R.S.C. values in the marginal range. These waters can be used for irrigation. In other areas, R.S.C. values fall within the not suitable condition.

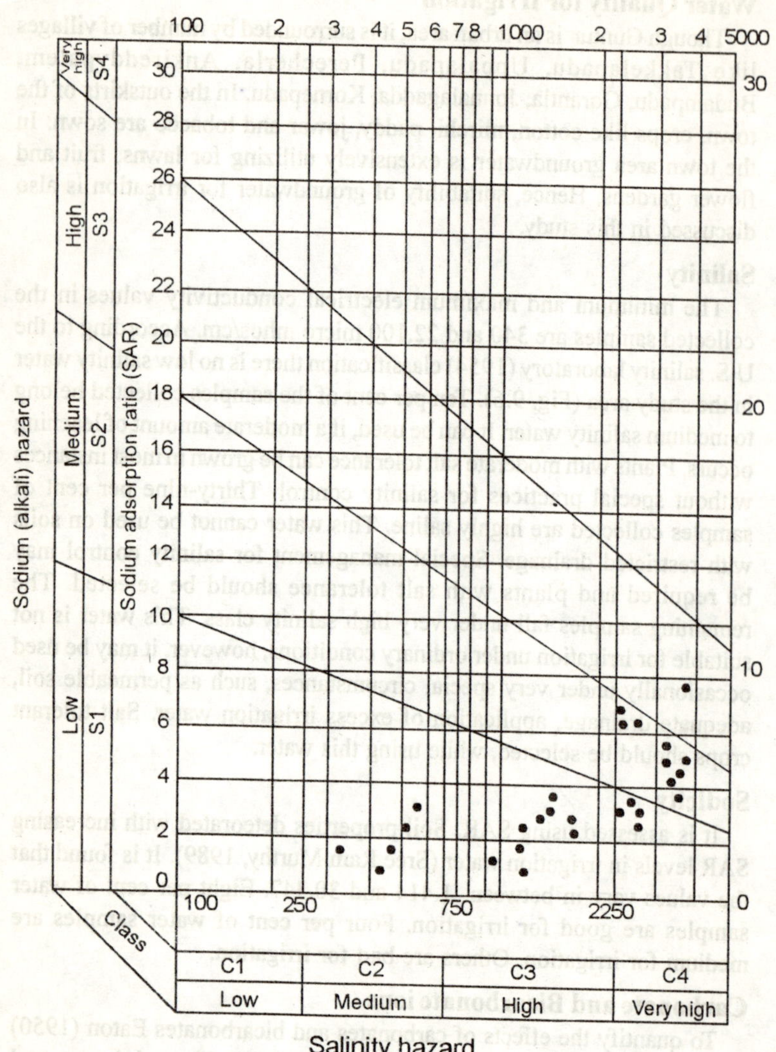

Fig. 9.6 Classification of water for agricultural use after Wilcox 1955

Magnesium hazard

Szabolcs and Darab (1964) proposed Magnesium hazard (MH) for irrigation water and where MH>50 the effects are considered to be harmful.

MH value varies from 5.479 to 94.25 in the water samples. Thirty-six samples have the MH value greater than 50.

Water Quality for livestock and poultry

EC and TDS are useful guide lines to know the suitability of water for livestock and poultry (NASNAF, 1972).

Twenty-six per cent of water samples collected from the wells are excellent for all classes of livestock and poultry. Fifty-six per cent of water samples collected in the study area are very satisfactory for all classes of livestock and poultry. These waters may cause temporary and mild diarrhoea in livestock not accustomed to it, or water dropping in poultry. Four per cent of the samples are satisfactory for livestock, but may cause temporary diarrhoea or be refused at first by animals not accustomed to it. Poor quality of water for poultry, often causing water increased mortality and decreased growth especially in turkeys. Four per cent of water samples collected in the study area can be used with reasonable safety for dairy and beef cattle, sheep, swine and horses. It is better avoid to use this water for pregnant or lactating animals. This water is not acceptable for poultry. Four per cent of samples are unfit for poultry and probably for swine. There is a considerable risk in using for pregnant or lactating cows, horses, or sheep or for the young of these special. The remaining two per cent of the samples cannot be recommended for use under any conditions.

References

Brown, E., Snoudsted, M.W. and Fishmen, M.J. (1970) in techniques of water resources, investigations of US Geol. Surv. Book 5, US Govt. Printing Office, Washington DC.

Eaton, F.M. (1950) Significance of carbonates in irrigation waters, Soil Sci., 69, pp. 123-33.

Karant, K.R. (1989) Hydrogeology, Tata-McGraw Hill, New Delhi, 458 pp.

National Academy of Sciences and National Academy of Engineering (1972) Water quality criteria, on REP. EPA-R373-033. Washington.

Sankara Pitchaiah, P. and Rajendra Rao, K. (1998) Groundwater hydrochemistry from Guntur Municipal Corporation, Andhra Pradesh,

Inland water resources : India, Ed. M.K. Durga Prasad and P. Sankara Pitchaiah, Discovery Publishing House.

Sawyer, C.N. and Mc Carty, P.L. (1967) Chemistry for sanitary engineers 2nd ed., McGraw Hill Co., New York, 518 pp.

Sreerama Murthy, K. (1989) Effect of irrigation water quality on soil-rice inter-relationships, Annamalai Univ. Publ. Annamalai Nagar, 200 pp.

Szabolcs, I. and Darab, C. (1964) The influence of irrigation water of high sodium carbonate content soils. Proc. 8th International Congress of ISSS. Trans. ii, pp. 803-12.

US Salinity Laboratory (1954) Diagnosis and improvement of saline and alkaline soils, US Dept. Agric. Handbook, 60, 160 pp.

World Health Organization (1978) Nitrates, nitrites and N-nitro compounds, Geneva, Environmental Health Criteria 5.

10

Groundwater Hydrochemistry from Guntur Municipal Corporation, Andhra Pradesh

P. Sankara Pitchaiah
K. Rajendra Rao

Introduction

Chemical composition of groundwater depends on many factors. Various physical and chemical properties reflect the hydrochemistry. In view of the utility of the groundwater in the Guntur Municipal Corporation (Fig. 10.1) present investigation is aimed at the study of hydrochemistry of these waters.

Methodology

Groundwater samples (49 nos.) are collected from the GMC area and its surroundings. The samples are analysed for various chemical constituents (Rajendra Rao and Sankara Pitchaiah, 1998). The results are plotted in the Durov's expanded diagram (Fig. 10.2) and the involved hydrochemical processes are discussed (Table 10.1). Relative frequency diagrams are prepared (Fig. 10.3) and the observations are presented.

Hydrochemistry

Earlier, Piper's trilinear diagram (1944, 1953) was in use to explain hydrochemistry of water. In view of the several drawbacks, Durov's expanded diagram (Lloyd and Heathcote, 1985) is adopted in the present investigation.

Expanded Durov Diagram

An Expanded Durov Diagram is prepared to discuss the water types and associated processes in the study area. Based on the diagram, the groundwater is classified into six groups (Table 10.1). These groups and per cent of samples fall in each group suggest that the groundwater of the GMC has undergone ion exchange, dissolution and mixing processes.

Ion-Concentration Percentage–Frequency Relationship

These graphs are proposed as supplements for the conventional trilinear diagrams (Zekaisen and Aldhakeel, 1986).

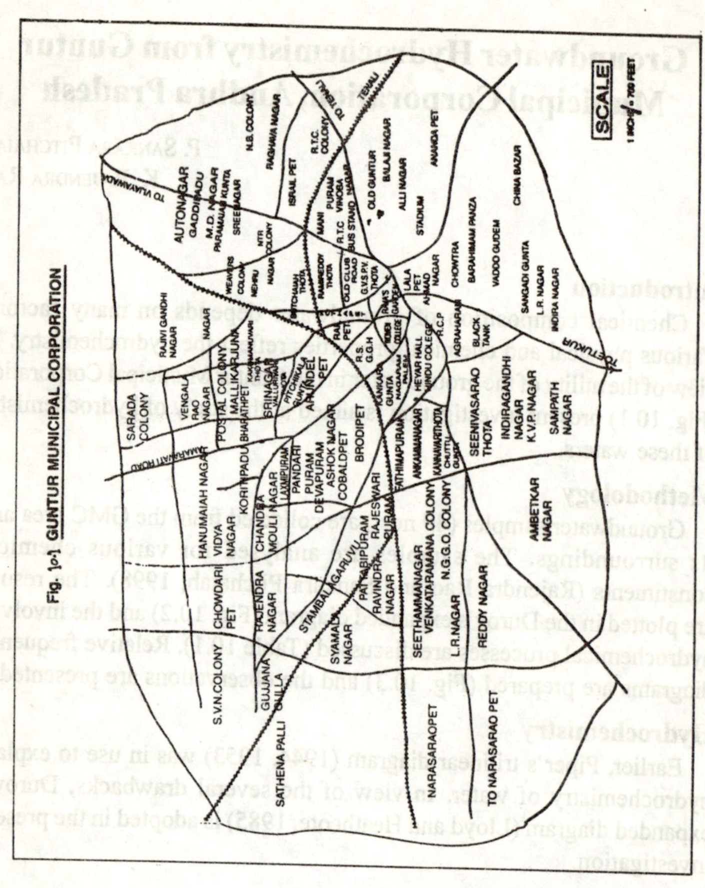

Fig. 10.1 GUNTUR MUNICIPAL CORPORATION

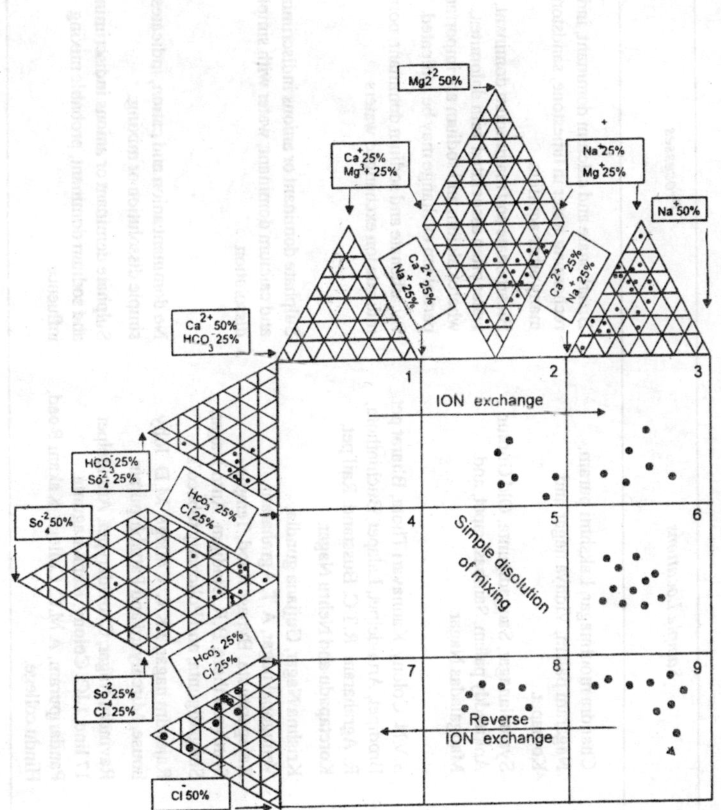

Fig.10.2 Expanded Durov Diagram

Table 10.1 : Water types and chemical processes involved

Sl. No.	*Water type*	*Per cent of samples*	*Sample Locations*	*Processes*
1.	Group–I	10	Chandra moulinagar, Lakshmi puram, Nagaram palem, Vidhya nagar and Kothapet.	Bicarbonate and calcium dominant, indicates recharging water in limestone, sandstone and many other aquifers.
2.	Group–II	12	Syamalanagar, Sangadigunta, Old Guntur, Ankireddy palem, Sambasivapet, and Mangaladas Nagar.	Bicarbonate and magnesium dominant, water often associated with dolomites, where calcium and sodium are important partial ion exchange may be indicated.
3.	Group–III	24	S.V.N. Colony, Kannavari Thota, Bharat pet, Brodipet, Arundelpet, Lalapet, Buchiathota, R. Agraharam, R.T.C. Busstand, Rail pet, Koretapadu and Nehru Nagar.	Bicarbonate and sodium dominant, normally indicate ion exchanged waters
4.	Group–IV	24	Krishna Nagar, Gujjana gundla, Venugopal nagar, A.T. Agraharam, Rajagarithota, Bypass Raod, Temple, Srinivasa thota, B.R. Stadium, Auto Nagar, Sangadigunta, and Polichetty tobaco's.	Sulphate dominant or anions indiscriminant and calcium dominant, water with simple dissolution
5.	Group–V	10	Rajendra nagar, J.K.C. College, M.D. Johy house, Mirchiyard and Pattabhipuram.	No dominant anion and cation, indicates simple dissolution or mixing.
6.	Group–VI	20	Ravindra Nagar, Chowdarypet, Arundelpet 17 line, LIC. Colony, Kankaragunta, Pandaripuram, A.M.C. College, Kakani Road, Hindu college.	Sulphate dominant or anions indiscriminant, and sodium dominant, probable mixing influence

E.C.

Most of the samples have Relative Frequency (R.F.) within the range 0-4. However, there is no change from premonsoon to post monsoon (Fig. 10.3a).

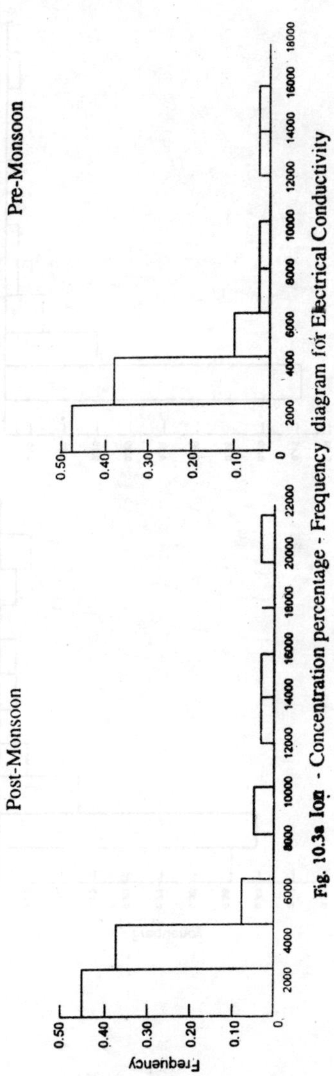

Fig. 10.3a Ion - Concentration percentage - Frequency diagram for Electrical Conductivity

T.D.S.

Most of the samples have R.F. within the range 0-3,000. There is a change in R.F. about 0.02 from premonsoon to post monsoon (Fig. 10.3b).

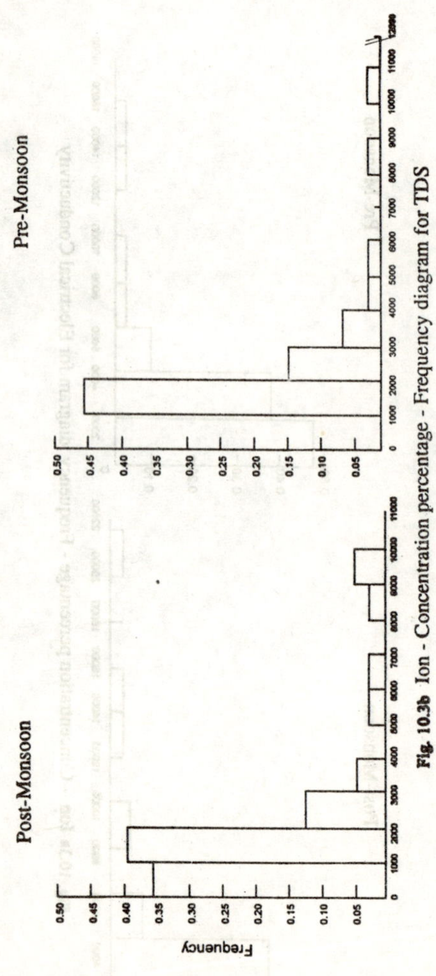

Fig. 10.3b Ion - Concentration percentage - Frequency diagram for TDS

Hardness

Almost all the samples have R.F. within the range 0-2,000. It is observed that there is a change in R.F. about 0.09 from premonsoon to post monsoon (Fig. 10.3c).

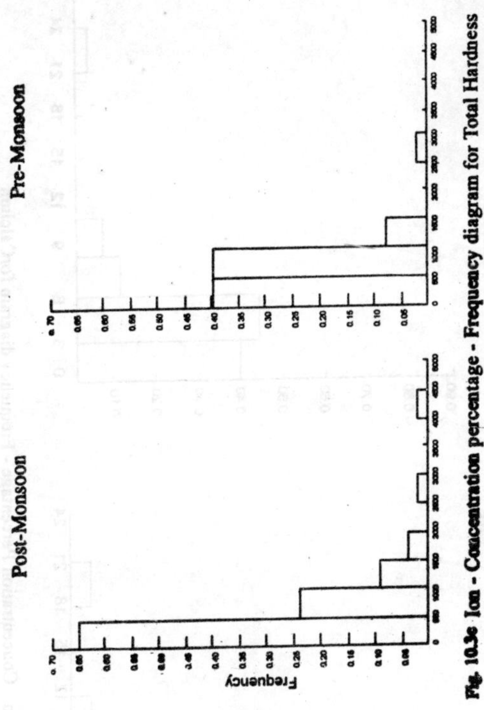

Fig. 10.3c Ion - Concentration percentage - Frequency diagram for Total Hardness

Calcium

Almost all the samples have R.F. within the range 0-6. From premonsoon to postmonsoon, R.F. of Calcium changes slightly (0.02) (Fig. 10.3d).

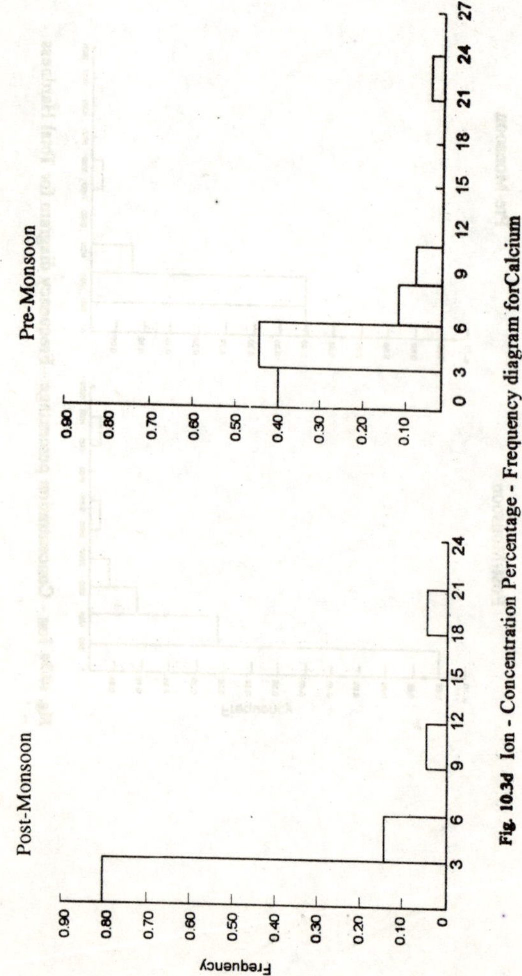

Fig. 10.3d Ion - Concentration Percentage - Frequency diagram forCalcium

Magnesium

Most of the samples have R.F. within the range 0-40. There is no considerable change in R.F. from premonsoon to post monsoon (Fig. 10.3e).

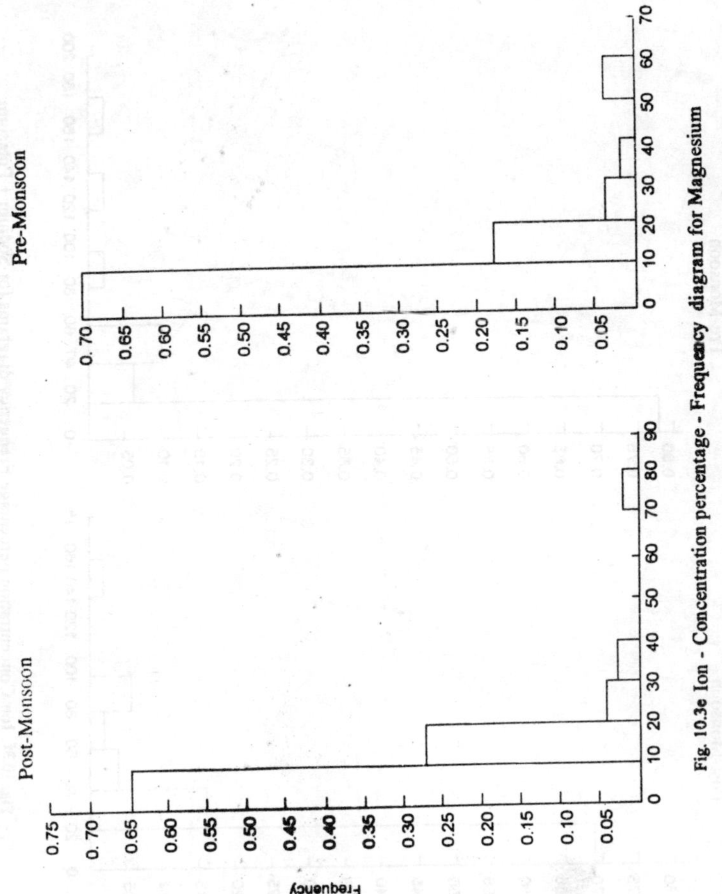

Fig. 10.3e Ion - Concentration percentage - Frequency diagram for Magnesium

Sodium and Potassium

Most of the samples have R.F. within 0-10 range, but there is no change from premonsoon to post monsoon (Fig. 10.3f).

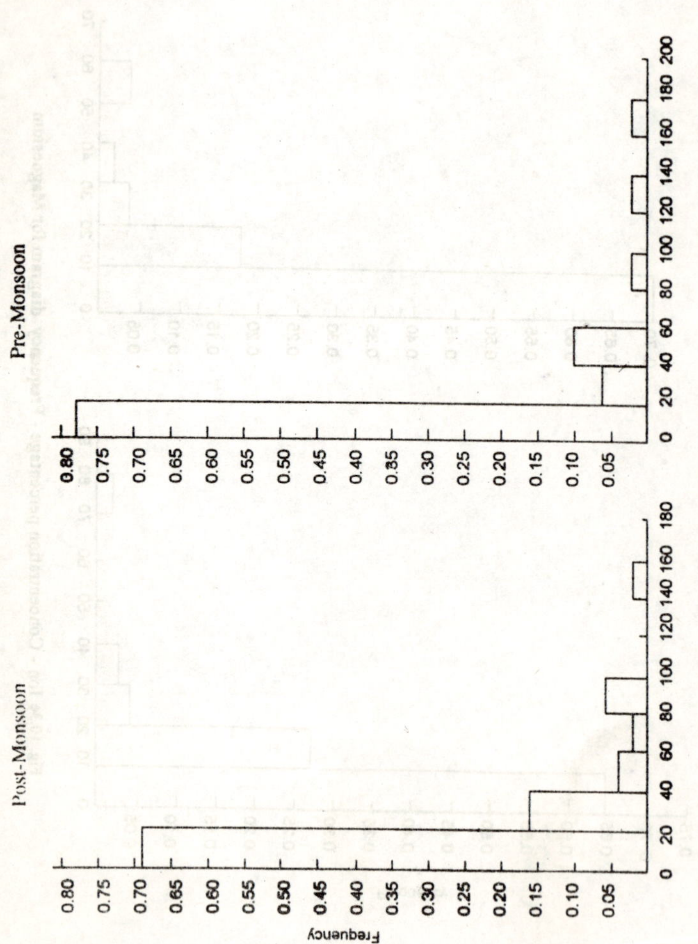

Fig. 10.3f **Ion Concentration percentage Frequency diagrams for Sodium + Potasium**

Bicarbonates

Widespread range (2 to 12) is observed for Bicarbonates. There is a little change in R.F. about 0.02 from premonsoon to post monsoon (Fig. 10.3g).

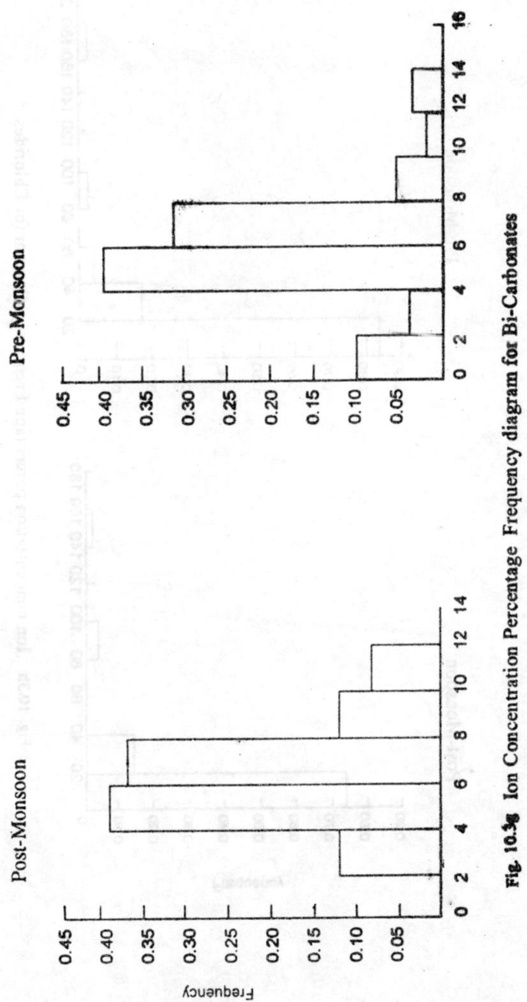

Fig. 10.3g Ion Concentration Percentage Frequency diagram for Bi-Carbonates

Chlorides

Most of the samples have R.F. within the range 0-40. From premonsoon to post monsoon 0.1 change in R.F. is noticed (Fig. 10.3h).

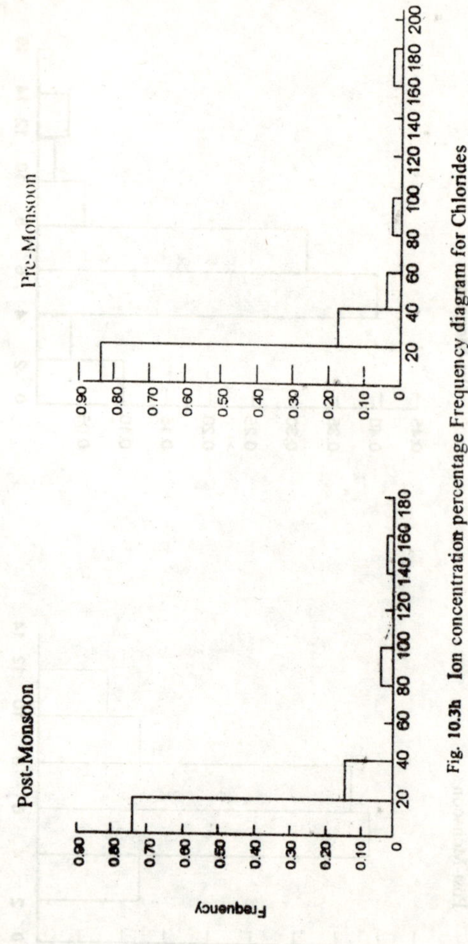

Fig. 10.3h **Ion concentration percentage Frequency diagram for Chlorides**

Sulphates

Most of the samples have R.F. within the range 2-10. Seasonal change is negligible (Fig. 10.3i).

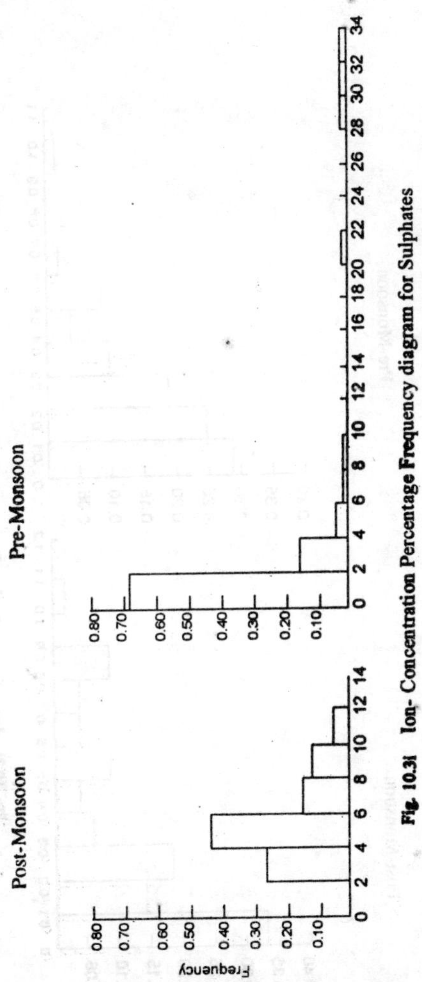

Fig. 10.3i Ion- Concentration Percentage Frequency diagram for Sulphates

Nitrates

Very widespread range (0-9) is observed for nitrate. Monsoonal change is about 0.08 (Fig. 10.3j).

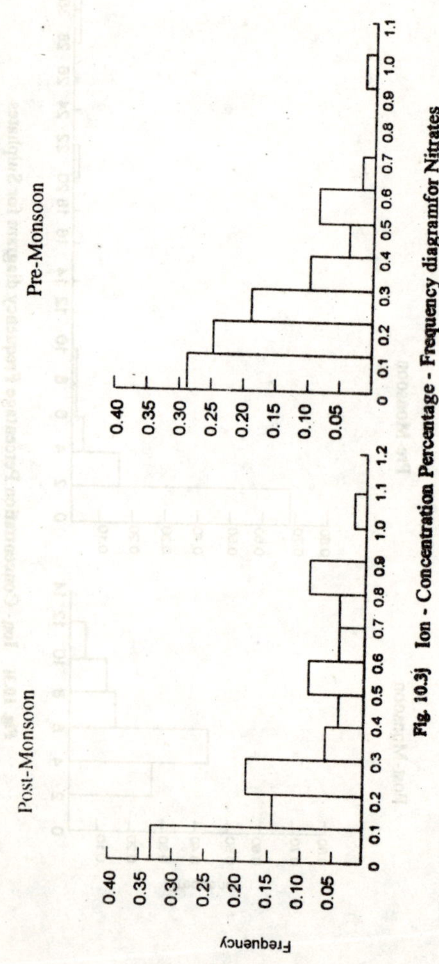

Fig. 10.3j Ion - Concentration Percentage - Frequency diagramfor Nitrates

References

Lloyd, J.W. and Heathcote, J.A.a. (1985) Natural inorganic hydrochemistry in relation to groundwater-An introduction, Clarendon Press, Oxford, 296 p.

Piper, A.M., (1944) A graphic procedure in the geochemical interpretation of water Analysis, Am. Geophys. Union. Trans. 25, pp. 914-23.

Piper, A.M. (1953) A graphic procedure in the geochemical interpretation of water Analysis, U.S. Geol. Surv. Groundwater Note 12.

Rajendra Rao, K. and Sankara Pitchaiah, P. (1998) Groundwater quality from Guntur Municipal Corporation, Andhra Pradesh, Inland water resources : India, Ed. M.K. Durga Prasad and P. Sankara Pitchaiah.

Zakai Sen and Aldhakeel, A. (1986) Hydro-chemical facies evaluation in Umm Er Radhuma limestone, Eastern Saudi Arabia, Groundwater, 24 (5), pp. 626-35.

11

A Study on the Water Quality, Plankton and Fish Yield in a Carp Culture Pond in Guntur District, Andhra Pradesh

P. PADMAVATHI,
M.K. DURGA PRASAD
AND
P. AJITHA

Introduction

The studies on fish production in relation to water quality, plankton, fertilizers and supplementary food in fish ponds of India are very meagre. Among the available literature, many papers deal on the hydrology and plankton. Mention may be made to the works of Ganapathi and Chacko (1951), Chacko and Krishnamurthy (1954), Alikunhi *et al.* (1955), Das and Srivastava (1956), George (1962 and 1966), Michael (1964 and 1968), Sreenivasan (1964 a & b), Zafar (1964 and 1967), Moitra and Bhattacharya (1965), Khan and Siddiqui (1971), Moitra and Mukherjee (1972), Ghosh and Hanuamantha Rao (1974), Mandal (1976), Jana *et al.* (1980), and Bose (Thesis unpubl. 1985). The lower deltaic region of river Krishna, south India supports a number of carp culture ponds extending an area of about 40,000 hectares. So far, no attempt has been made to study the impact of various parameters on the fish yield of this area. Hence, the present investigation was undertaken to make a comprehensive study on the impact of physico-chemical parameters of water, plankton, fertilizers and supplementary food on the fish production in a carp culture pond at Kancherlapalem.

Study Area

The pond under investigation is situated in a village, Kancherlapalem at a distance of 6 km north of Tenali (16°15'N and 80°35'E), Guntur district, Andhra Pradesh, India. The extent of the pond is 20,234 sq. m. (2 hectares) and it is mainly used for the culture of Indian major carps. The depth of the pond ranged between 0.75 m in March 1987 and 2m in September 1986. The pond is seasonal and dries up towards midsummer. It derives its water supply mainly from the Nizampatnam irrigation canal of river Krishna.

Materials and Methods

Physico-chemical and biological parameters of water were studied from September 1986 to March 1987 and the collections were made around 9 A.M. The physico-chemical parameters studied were the temperature (air and surface water), transparency, rainfall, dissolved oxygen, pH, total alkalinity, total hardness, nitrates, nitrites and orthophosphates. The methods adopted for the estimation of various parameters were the standard methods suggested by Welch (1948), Golterman and Clymo (1969) and APHA (1975). Rainfall data was collected from Duggirala meteorological Observatory, near Tenali. The biological parameters studied were plankton and fish. Plankton samples were collected by filtering 100 litres of water through a plankton net made up of silk bolting cloth No. 25 and the plankton obtained were fixed in 5% formalin. Enumeration of the individual plankters was done on the lines recommended by Edmondson (1966). The zooplankters were identified up to species while the phytoplankters were identified up to generic level. Carp fingerlings were stocked in the month of September. Prior to the release of the fingerlings, fertilization of the pond was taken up with 7.5 tonnes of cowdung and 600 kg of lime. Fertilization and liming were done once in every two months with 2 tonnes of cowdung and 200kg of lime. In addition to the natural food, the fish were fed daily with supplementary feed of groundnut oilcake, rice bran and fish meal in the ratio of 2:2:1 at the rate of 2% body weight and in February and March at the rate of 5% body weight. The fish were sampled by using cast net every month to see the growth rate. The data on the gut analysis of *Catla catla, Labeo rohita and Cirrhinus mrigala* studied by Alikunhi (1957), Natarajan and Jhingran (1961), Chakrabarty and Singh (1963), and Khan and Siddiqui (1973) were considered and discussed for arriving at some conclusions.

Results

PHYSICO-CHEMICAL PARAMETERS : The results of the parameters studied are given in Table 11.1.

Plankton : Plankton studies during this period revealed that phytoplankton was more abundant than the zooplankton.

Phytoplankton : Twenty-five genera of phytoplankters belonging to Myxophyceae, Chlorophyceae, Baccillariophyceae, Dinophyceae and Euglenineae were identified. The numerical abundance and seasonal variations are given in Table 11.2. It was observed that the genera of Myxophyceae dominated the other groups and all the groups present

Table 11.1 : Physico-Chemical Parameters of Kancherlapalem Fish Pond

Month & Year	*Air temp.* °C	*Surface water temp.* °C	*Water trans-parency cm*	*Total rain-fall mm*	*Dissolved oxygen mg/l*	*pH*	*Total alkal-inity mg/l as $CaCO_3$*	*Total hard-ness mg/l as $CaCO_3$*	*Phos-phates μg/l*	*Nitrite μg/l*	*Nitrates μg/l*
Sep. 1986	33.0	31.0	25.0	97.3	4.8	9.00	*	*	*	*	*
Oct. 1986	31.5	31.5	40.0	56.0	4.0	9.00	240.0	125.0	99.20	5.5	34.5
Nov. 1986	29.0	29.0	30.0	73.0	3.2	8.50	234.0	115.0	106.25	8.8	48.96
Dec. 1986	29.0	28.0	32.5	0.0	2.6	8.36	248.0	147.5	75.00	15.55	62.24
Jan. 1987	28.3	27.8	26.5	0.0	3.9	9.01	254.0	170.0	90.62	32.22	82.95
Feb. 1987	30.0	28.0	28.0	0.0	3.4	8.78	243.0	117.5	97.00	17.77	74.09
Mar. 1987	32.0	30.0	18.0	60.0	3.0	8.54	288.0	135.0	112.0	36.66	94.94

Note : *Not estimated

Table 11.2 : Seasonal Abundance of Phytoplankton in Kancherlapalem Fish Pond

Phytoplankton	*Sep. 1986*	*Oct. 1986*	*Nov. 1986*	*Dec. 1986*	*Jan. 1987*	*Feb. 1987*	*Mar. 1987*
Myxophyceae							
Oscillatoria	23,325	3,563	720	80	664	593	750
Anabaena	5,500	2,588	135	417	321	1,510	100
Lyngbya	73	-	135	-	82	-	-
Raphidiopsis	-	1,350	-	60	-	-	-
Cylindrospermum	525	50	180	45	-	-	-
Arthrospira	743	338	-	100	4,757	1,513	50
Microcystis	1,875	150	270	135	-	290	-
Merismopaedia	-	35	-	73	64	888	300
Total	32,041	8,075	1,440	910	5,888	4,794	1,200
Percentage in Total	98.96%	93.10%	59.77%	53.69%	60.22%	78.25%	54.55%
Chlorophyceae							
Pediastrum	4	75	11	61	29	-	-
Scenedesmus	101	75	270	55	-	493	100
Ankistrodesmus	-	-	-	35	-	-	100
Micractinium	-	-	135	-	-	-	-
Tetraedon	-	35	-	-	64	-	-
Pandorina	-	-	-	-	-	-	35
Closterium	75	113	13	15	-	63	-
Cosmarium	-	-	-	-	57	-	-
Total	190	298	429	166	150	556	250
Percentage in Total	0.58%	3.43%	17.80%	9.79%	1.88%	7.40%	11.36%
Bacillariophyceae							
Melosira	-	35	-	-	-	-	-
Synedra	90	113	180	128	168	148	-
Navicula	20	-	-	45	35	43	-
Amphora	-	-	-	-	-	-	50
Cymbella	-	-	-	35	-	-	-
Diatoma	-	-	-	-	-	-	50
Total	110	148	180	208	230	191	100
Percentage in Total	0.33%	1.70%	7.48%	12.27%	2.88%	2.54%	4.54%

(Contd.)

Table 11.2 (Contd.)

Phytoplankton	*Sep. 1986*	*Oct. 1986*	*Nov. 1986*	*Dec. 1987*	*Jan. 1987*	*Feb. 1987*	*Mar. 1987*
Englenieae							
Euglena	27	125	-	270	1,809	552	500
Phacuas	8	25	-	-	401	49	-
Total	35	150	-	270	2,210	601	500
Percentage in total	0.10%	1.70%	-	15.93%	27.76%	8.0%	22.73%
Dinophyceae							
Peridinium	-	-	360	141	576	278	150
Percentage in total	-	-	14.95%	8.32%	7.24%	3.70%	6.82%
Total Phyto-plankton	32,376	8,670	2,409	1,695	9,054	6,420	2,200

The values are represented as number of individuals per litre.

have normally two maxima except the Bacillariophyceae which has a single peak. The dominating group, Myxophyceae has a primary peak in September and a secondary peak in January. Chlorophyceae showed a major peak in February and a minor one in November, Euglenieae has major peak in January and a minor in October, Dinophyceae showed a major peak in January and a minor one in November whereas Bacillariophyceae has a single peak in January. Since Myxophyceae dominated all other groups, its abundance reflected in the major peak of the phytoplankton in September, touching a minimum in December with a minor peak in the month of January.

Zooplankton : Qualitative and quantitative studies indicated that rotifers were the most abundant and dominating group of zooplankton followed by copepods and cladocerans. Other groups such as protozoans, ostracods and insect larvae were also noticed in meagre numbers and they are referred to as miscellaneous forms. Among the zooplankters, 17 genera were identified, of which, the dominating genus was *Brachionus*. The zooplankton has two peaks, a major one in September and a minor one in January. The quantitative and seasonal abundance of zooplankters are given in Table 11.3.

The dominant group, Rotifera was represented by 17 species belonging to 8 genera (Table 11.3). Among them, the genus *Brachionus* was common and represented by 7 species. The primary zooplankton peak was due to the high density of *Keratella tropica* in September. The secondary peak was due to the abundance of *Asplanchna intermedia and Brachionus* spp. Next in abundance was the naupliar and copepodid stages. These stages have also contributed to the primary peak and thereafter there was a gradual decline. The copepods were represented by 4 genera. The calanoid copepods were represented by 3 species belonging to 2 genera, namely, *Heliodiaptomus and Rhinediaptomus* and were represented by a few numbers in the months of September, October and February, whereas the cyclopoids were represented by 2 species belonging to 2 genera, namely, *Mesocyclops and Microcyclops* and they were present more or less throughout the period of study. Their significant decline was noticed in March. Next in abundance was the cladocerans represented by 6 species belonging to 5 genera (Table 11.3). Their major peak was seen in September followed by a minor peak in January.

Fish : Fingerlings of 4 species of carp, viz., *Catla catla, Labeo rohita, Cirrhinus mrigala and Ctenopharyngodon idella* were introduced in September 1986. A total number of 9,800 fingerlings, of which 6,000 *Catla catla,* 3,000 *Labeo rohita,* 750 *Cirrhinus mrigala* and 50 *Ctenopharyngodon idella* were stocked. The average weights of *C. catla catla, L. rohita, C. mrigala and C. idella* were 5.0, 7.3, 5.0, and 10.0g respectively. The monthly average weight and length of three major carps, number of fish harvested and their total weights are given in Table 11.4.

Discussion

A number of interdependent physical, chemical and biological factors influenced the seasonal succession of the plankton in Kancherlapalem pond which in turn effected the fish yield. The incoming canal water, the rainfall, the added fertilizers, the supplementary feed, and the metabolites of the organisms effected the water quality and fish production in the pond.

During the period of observation, the temperature of the surface water was not subjected to a significant variation. Such a narrow range of variation in temperature is characteristic of tropical waters (Jana, 1979). The fluctuations of the water temperature did not show any definite influence on the total plankton. Similar observations were also made by Jana (1973) and Bose (1985). Water transparency reflected the suspended matter and plankton abundance in the water. During the present

Table 11.3 : Seasonal Abundance of Zooplankton in Kancherlapalem Fish Pond

Phytoplankton	*Sep. 1986*	*Oct. 1986*	*Nov. 1986*	*Dec. 1986*	*Jan. 1987*	*Feb. 1987*	*Mar. 1987*
Rotifera							
Brachionus calyciflorus	21	7	5	2	6	11	10
B. angularis	25	68	8	3	38	7	–
B. forficula	4	104	23	19	12	2	–
B. falcatus	–	2	–	–	–	–	–
B. diversicornis	4	154	36	11	54	46	61
B. quadridentatus	–	–	–	2	4	–	3
Keratella tropica	2,302	7	49	3	–	3	–
Asplanchna intermedia	14	23	65	21	80	9	11
Trichocerca sp.	3	5	6	4	4	2	–
Lecane spp.	8	20	2	1	7	2	–
Filinia longiseta	26	5	2	2	4	2	–
F. pejleri	59	5	1	–	–	–	–
Ployarthra remata	–	5	12	3	5	5	–
Hexathra sp.	–	2	2	2	–	2	–
Total	2,446	424	219	75	208	93	91
Percentage in Total	83.42%	77.80%	51.90%	70.09%	81.89%	76.20%	77.13%
Cladocera							
Diaphanosoma excisum	9	3	2	4	11	2	6
Ceriodaphnia conuta	5	–	–	–	–	–	–
Moina spp.	4	–	2	3	4	1	4
Macrothrix spinosa	–	–	–	1	–	1	–
Alona pulchella	–	–	–	1	1	1	–
Total	18	3	4	9	16	5	10
Percentage in Total	0.61%	0.55%	0.95%	8.41%	6.30%	4.00%	8.47%

Table 11.3 (Contd.)

Phytoplankton	*Sep 1986*	*Oct 1986*	*Nov 1986*	*Dec 1987*	*Jan 1987*	*Feb 1987*	*Mar 1987*
Copepoda							
Mesocyclops thermocyclopoides	9	3	1	1	2	4	1
Microcyclops varicans	27	21	37	3	5	3	1
Heiodiaptomus cinctus	2	–	–	–	–	2	–
H. viduus	5	1	–	–	–	1	–
Rhinediaptomus indicus	4	–	–	–	–	–	–
Total	47	25	38	4	7	10	2
Percentage in Total	1.59%	4.59%	9.00%	3.74%	2.76%	8.00%	1.69%
Crustacean Larvae	413	90	154	10	11	14	10
Percentage in Total	13.97%	16.51%	36.49%	9.35%	4.33%	11.50%	8.47%
Miscellaneous	12	3	7	9	12	–	5
Percentage in total	0.41%	0.55%	1.66%	8.41%	4.72%	–	4.24%
Total Zooplankton	2,956	545	422	107	254	122	118

The values are represented as number of individuals per litre.

observation, low transparency values were noticed in September and January during which period phyto and zooplankton were abundant. Similar observation was made earlier by Michael (1969). Indirect relationship between secchi disc visibility and phytoplankton was also observed by Almazan & Boyd (1978). The low transparency observed during March was due to the mixing of bottom mud in pond water by wind action and active movements of the fish. The fluctuations in pH were not marked. The range being 8.36 to 9.01. High pH values recorded in September and January coincided with the abundant phytoplankton indicated that higher photosynthetic activity would increase the pH value. This observation is in confirmity with the findings of Das (1961), Moitra

Table 11.4 : Monthly Average Weights, Lengths and Final Yield of Three Cultured Carps

Month & Year	*Catla*			*Rohu*			*Mrigal*		
	Number of fish sampled	*Weight* *g*	*Length* *cm*	*Number of fish sampled*	*Weight* *g*	*Length* *cm*	*Number of fish sampled*	*Weight* *g*	*Length* *cm*
Sep. 1986	20	5.0 ± 0.7	7.4 ± 0.64	20	7.3 ± 0.9	8.8 ± 0.76	20	5.0 ± 0.7	8.5 ± 0.60
Oct 1986	16	166.0 ± 23.9	20.3 ± 1.78	20	152.0 ± 40.6	21.2 ± 3.10	8	115.0 ± 34.0	20.3 ± 1.80
Nov 1986	20	201.0 ± 25.9	22.1 ± 1.41	18	234.0 ± 30.8	25.1 ± 1.37	20	166.5 ± 42.3	22.0 ± 2.10
Dec 1986	20	222.2 ± 34.8	23.4 ± 1.80	20	295.5 ± 46.3	26.3 ± 2.10	16	261.0 ± 62.1	25.5 ± 2.90
Jan 1987	20	255.0 ± 39.8	24.3 ± 2.45	10	320.0 ± 58.3	26.9 ± 3.40	–	–	–
Feb 1987	18	280.6 ± 54.4	26.7 ± 2.60	18	360.0 ± 39.4	29.1 ± 2.20	8	307.5 ± 60.1	28.8 ± 2.60

Mar. 1987	12	345.0 ± 37.9	30.1 ± 1.92	12	419.0 ± 59.0	32.5 ± 2.80	–	–	–
Apr. 1987	20	440.0 ± 57.9	32.0 ± 2.70	20	540.0 ± 81.8	38.5 ± 4.21	20	370.0 ± 66.9	35.4 ± 3.80

		Catla	Rohu	Mrigal	Grass Carp	Total
No. of fingerlings introduced	= 6,000	3,000	750	50	=	9,800
No. of fishes harvested	= 4,736	3,141	2,430	31	=	10,338
Total biomass of fish harvested (kgs)	= 2,096	1,699	905	17	=	4,717

Yield = 4.717 tonnes/2ha/207 days or 4.158 tonnes/ha/year.

and Bhattacharya (1965). The present results do not indicate any relationship with the abundance of zooplankton. Dissolved oxygen varied from a minimum of 2.6 mg/l in December to a maximum of 4.8 mg/l in September. Dissolved oxygen showed a direct relation to phytoplankton abundance because they directly involve in the photosynthetic activity and release oxygen. The total alkalinity and hardness did not show any relation to the total plankton. The total alkalinity and hardness decrease in the months of heavy rainfall and increase with the intense evaporation during summer months. Such observations were also made by Tucker (1958) and Hazelwood and Parker (1961). Nitrates and nitrites did not show any relationship with the plankton. Concentrations of these nutrients were constantly changing with the regular fertilization and supplementary feed given to the pond. Phosphates were observed to be high in the rainfall months, thereby indicating that this nutrient was obtained through the percolated waters from the neighbouring agricultural lands. Similar observations were also made by Michael (1969). When the phosphates were low in the month of December, the density of plankton was at its minimum.

The phytoplankton was dense during the months of September, October and January. In these months, a significant increase in the density of zooplankton was observed. This indicates that the phytoplankton supports the zooplankton production. The fluctuations of the Myxophyceae was closely followed by those of the rotifiers, indicating a positive relationship between these two groups. Wadajo and Belay (1984) also observed a direct relationship between the rotifers and Myxophyceae and between the Chlorophyceae and cladocerans. However, in the present study, no marked relation was observed between the chlorophyceae and cladocerans. This may be due to the fact that the members of the Chlorophyceae form the preferential food for the cultured fish, *Labeo rohita* (Khan and Siddiqui, 1973). In the present study, zooplankters were dominated by rotifers followed by copepods and cladocerans. In ponds of this area, where fish were not cultured, copepods and cladocerans were dominant over rotifers (Ranga Reddy, 1977, Thesis unpub.; Durga Prasad, 1981, Thesis unpubl.). The low number of cladocerans and copepods compared to rotifers in Kancherlapalem pond might be due to the preferential predation by the cultured fish. Khan and Siddiqui (1973) also reported that *Catla catla* feeds preferentially on crustacean zooplankton rather than rotifers and other planktonic organisms. Further, the rotifer abundance may be due to their high natality compared to copepods and cladocerans at optimal food conditions (Edmondson, 1965). The maximum number of cladocerans and copepods in September compared to other months was due to the hatching up of the resting eggs in monsoon season when the pond basin was filled with water. Similar observations were also made by Durga Prasad (Thesis unpubl., 1981) in a seasonal hillside pond at Vengalayapalem. Further, the initial fertilization given to the pond in August might be responsible in building up of cladoceran and copepod populations.

Carp fingerlings were stocked in September. In the months of October, the fingerlings of *Catla catla, Labeo rohita and Cirrhinus mrigala* attained average weights of 166, 152 and 115g measuring 20.3, 21.2 and 20.3 cms respectively. The weight gained by these fingerlings in this period indicated an initial spurt in their growth. This is due to the fact that these fingerlings are feeding on the abundant plankton (35,332 individuals/l) available in the pond. Detailed studies on gut analysis of various life stages of Indian major carps (Mookerjee, 1944-45; Alikunhi, 1957; Khan and Siddiqui, 1973) revealed that the fingerling stages of all the three carps feed preferably on zooplankton. The drastic reduction of zooplankton in the pond is initially due to the preferential predation of all the fingerlings. As the fingerlings grow to the adult stage, they change their mode of feeding. The adults of *Catla catla* feed preferentially on zooplankton (Natarajan & Jhingran, 1961; Khan and Siddiqui, 1973), *Labeo rohita* on phytoplankton (Alikunhi, 1957; Khan & Siddiqui, 1973), whereas *Cirrhinus mrigala* feeds on decayed organic matter and associated plankton (Alikunhi, 1957; Chakraborty and Singh, 1963; Khan and Siddiqui, 1973). As the populations of *C. catla and L. rohita* feed intensively on zoo- and phytoplankton respectively, the abundance of these plankton groups decreased markedly from October to March. The fertilization of the pond at different months just to supplement the nutrients for the growth of the plankters may not help for the recovery and sustenance of plankton in the pond. This is due to the fact that both phyto- and zooplankters whose production is influenced by the enrichment of nutrients were continuously fed by carps. Further, the supplementary feed given to the fish may not be sufficient to sustain the growing fish populations. As the fish grows, it requires more absolute amounts of food in order to sustain its potential growth and maintain its body (Winberg, 1956). In the present study, the fish were fed daily with supplementary feed only at the rate of 2% body weight to the number of fish actually stocked rather than the actual number of fish existing in the pond. This is due to the fact that some fingerlings of rohu and mrigal entered the pond through the incoming canal waters. The low density of plankton and insufficient quantities of supplementary feed given to the fish whose number is further increased by the entry of some carps through the incoming canal waters in this culture pond was ultimately responsible for their reduction in growth rate. The improved growth rate observed in the months of March and April may be due to the increase in the supply of supplementary feed (@ 5% body weight of fish). Thus in this culture pond, within a period of seven months (207 days), *C. catla* attained an average weight of 440g. The number of *catla* stocked was 6,000 whereas only 4,736 individuals were recovered with a biomass of 2,096 kg. The number of *L. rohita* stocked was 3,000 and the number recovered was 3,141. The average weight attained by them was 540g thus obtaining a biomass of 1,699 kg. In case of *C. mrigala*, only 750 fingerlings were introduced whereas 2,430 fishes have been harvested with an average weight of 370 g. The biomass obtained was 905 kg. The recovery of fish in case of *C. idella* was 60% and reached a biomass of 17 kg. It was observed that the number of *L. rohita and C. mrigala* harvested were more than the number introduced in the pond. This clearly indicates a significant number of these fishes entered the pond through the irrigational canal and colonized in the pond with the stocked fish. The total biomass produced in 2 hectare pond in 207 days (c. 7 months) was 4,717 kg, thus producing an average yield of 2.358 tonnes/ha/207days which corresponds to the production of 4.148 tonnes/ha/year.

However, better yield could be expected in Kancherlapalem fish pond by

1. Stocking the pond with optimum number of fish: Since the pond was stocked with 4,900 fish/ha with the combination of *Catla catla, Labeo rohita, Cirrhinus mrigala and Ctenopharyngodon idella* at the ratio of 3000:1500:375:25 respectively, better results were achieved. However, for better fish production, c.3000 fish/ha with 3:6:1 ratio of *C. catla, L. rohita and C. mrigala* may be introduced.

2. Preventing the intrusion of exogenous fish from the inflowing canal waters into the pond: Since the pond experienced the entry of exogenous fish including carps, the density of fish was further increased and this condition has led to poor production. However, for the better yield of fish, the entry of these fishes should be prevented by arranging proper meshes or screening sleeves at the inlet pipe or by adapting double filtration method. In any case, when these fishes entered and colonized in the pond, they should be removed by repeated nettings.

3. Maintaining good water quality: Since the water quality with respect to dissolved oxygen, total alkalinity, total hardness and nutrient levels was not up to the mark, these parameters should be maintained at optimum levels to achieve better yield of fish. Dissolved oxygen should be maintained above 5mg/l and it can be accomplished by aerating the pond water. Nutrient levels can be increased by adding organic and inorganic fertilizers such that the nitrate and phosphate concentration should be maintained at 1 to 2mg/l. Large difference in concentrations between the total alkalinity and total hardness often leads to rise in pH during daytime and it results in poor growth of fish. So their concentration should be equalized by adding lime.

4. Adapting regular and proper fertilization with organic and inorganic fertilizers which impart sufficient nutrients to the water and enrich the plankton production: During the culture period, the fertilization given to the pond was inadequate. The pond was fertilized with only cowdung in alternate months. This may not provide sufficient nutrients to the water to enhance the plankton production. So, the pond should be fertilized with organic and inorganic fertilizers at regular fortnight or monthly intervals. The addition of organic and inorganic fertilizers in the pond at the rate of c.20 tonnes of poultry manure, c. 900 kg of single superphosphate, c. 600 kg. of Urea and c. 150 kg. of muriate of potash per hectare for the culture period may result in better fish production.

5. Giving supplementary feed at the rate of 5% body weight of the fish : Since the supplementary feed given to the fish at the rate of 2% body weight is insufficient, it is felt that the fish existing in the pond be given the supplementary food at the rate of 5% body weight to achieve better production.

Acknowledgements

The authors wish to express their gratitude and indebtedness to late Prof. Y. Radhakrishna, Director, Centre for Aquaculture Research & Education, Nagarjuna University, for his constant guidance and encouragement. Thanks are due to the authorities of Nagarjuna University for providing facilities to carry out this work.

References

Alikunhi, K.H., (1957) Fish culture in India. Farming Bull. Indian counc. Agric. Res., V.20: pp. 144.

Alikunhi, K.H., Chaudhri, H. and Ramachandran, V., (1955) On the mortality of carp fry in nursery ponds and the role of plankton in their survival and growth. Indian J. Fish., V. 2: pp. 257-313.

Almazan, G. and Boyd, C.E., (1978) An evaluation of sechi disc visibility for estimating density in fish ponds. Hydrobiologia, V. 61: pp. 205-08.

APHA (American Public Health Association), (1975) Standard methods for the Examination of Waters and Wastewaters, 14th ed. American Public Health Association, Washington, D.C. 1193 p.

Bose, R.S.C., (1985) Studies on zooplankton with special reference to Cladocera in a tropical fish farm near Vuyyuru, Krishna district (A.P., India). M. Phil. Thesis, Nagarjuna University.

Chacko, P.I. and Krishnamurthy, B., (1954) On the plankton of three freshwater fish ponds in Madras city, India. Indo-pacific Fish. Counc. (UNESCO): pp. 103-07.

Chakrabarty, R.D. and Singh, S.B., (1963) Observations on some aspects of the fishery biology of the mrigal, *Cirrhinus mrigala* (Ham.) from Allahabad. Indian J. fish., V. 10A: pp. 209-32.

Das, S.M., (1961) Hydrogen-ion concentration, plankton and fish in freshwater Eutrophic lakes of India. Nature, London, V. 191: pp. 511-12.

Das, S.M., and Srivastava, V.K. (1956) Quantitative studies on freshwater plankton. 1. Plankton of a fish tank in Lucknow, India. Proc. Nat. Acad. Sci. India, V. 26: pp. 85-92.

Durga Prasad, M.K., (1981) Studies on systematics and ecology of Branchiopods of Guntur district and its environs (A.P., India) Ph. D. Thesis, Nagarjuna University (Unpubl.).

Edmondson, W.T., (1965) Reproductive rate of planktonic rotifers as related to food and temperature in nature. Ecol. Mongor., V. 35: pp. 61-111.

Edmondson, W.T., (1966) Freshwater Biology. John Wiley & Sons, Inc., New York, 1248 p.

Ganapathi, S.V. and Chacko, P.I., (1951) Investigations on plankton production in ponds Chelput fish farm, Madras. The limnology of four ponds. Proc. Inst. Ass. Theo. Limn., V. 2: pp. 11-12.

George, M.G., (1962) Occurrence of a permanent algal bloom in a fish tank at Delhi with special reference to factors responsible for its production. Proc. Ind. Acad. Sci. V. 61 pp. 354-62.

George, M.G., (1966) Comparative plankton ecology of five fish tanks in Delhi, India. Hydrobiologia, V. 27: pp. 81-108.

Ghosh, A. and Hanumantha Rao, L., (1974) Studies on the hydrobiological conditions of a sewage-fed pond with a note on their role in fish culture. J. Inland Fish. Soc. India, V. 6: pp. 51-61.

Golterman, H.L. and Clymo, R.S., (1969) Methods for chemical analysis of freshwaters. IBP Hand Book No. 8:172 p.

Hazelwood, H.D. and Parker, R.A., (1961) Population dynamics of some freshwater zooplankton. Ecology, V. 42: pp. 266-74.

Jana, B.B., (1973) Seasonal periodicity of plankton in a freshwater pond in West Bengal, India. Int. Revue. ges. Hydrobiol., V. 58: pp. 127-43.

Jana, B.B., (1979) Temporal plankton succession and ecology of a tropical tank in West Bengal. Int. Revue. ges. Hydrobiol., V. 64: pp. 661-671.

Jana, B.B., De, U.K. and Das, R.N., (1980) Environmental factors affecting the seasonal changes of net plankton in two tropical fish ponds in India. Schweiz. Z. Hydrol., V. 42: pp. 225-46.

Khan, A.A. and Siddiqui, A.Q., (1971) Primary production in a tropical fish pond at Aligarh, India Hydrobiologia, V. 37: pp. 447-56.

Khan, R.A. and Siddiqui, A.Q., (1973) Food selection by *Labeo rohita* (Ham.) and its feeding relationship with other major carps. Hydrobiologia, V. 43: pp. 429-42.

Mandal, B.K., (1976) Studies on the primary productivity and physico-chemical factors of two fish ponds at Burdwan, W. Bengal (India). Acta. Hydrobiol., V. 18 : pp. 175-812.

Michael, R.G., (1964) Limnological investigations on pond plankton, macrofauna and chemical constituents of water and their bearing on fish production. Unpubl. Ph.D. Thesis, Calcutta University, India, 165 p.

Michael, R.G., (1968) Studies on the zooplankton of a tropical fish pond. Hydrobiologia, V. 32: pp. 47-68.

Michael, R.G., (1969) Seasonal trend in physico-chemical factors of a freshwater fish pond and their role in fish culture. Hydrobiologia, V. 33: pp. 144-60.

Moitra, S.K. and Bhattacharya, B.K., (1965) Some hydrobiological factors affecting plankton production in a fish pond in Kalyani, West Bengal, India. Ichthyyologica, V. 4 : pp. 8-12.

Motira, S.K. and Mukherjee, S.K., (1972) Studies on the freshwater plankton of a fish pond in Kalyani, West Bengal, India Vest. cesk. spol. zool., V. 36: pp. 23-28.

Natrajan, A.V. and Jhingran, A.G., (1961) Index preponderance-A method of grading the food elements in the stomach analysis of fishes Indian J. Fish., V. 8: pp. 54-59.

Ranga Reddy, Y., (1977) Studies on systematic and ecology of freeliving freshwater copepods of Guntur and its environs (Andhra Pradesh, India). Ph. D. Thesis, Nagarjuna University (Unpubl.)

Sreenivasan, A., (1964a) The limnology, primary production and fish production in a tropical pond. Limnol. Oceanogr., V. 9: pp. 391-96.

Sreenivasan, A., (!964b) Limnological studies and fish yield in three upland lakes of Madras State, India. Limnol. Oceanogr., V. 9 : pp. 564-75.

Tucker, D.S., (1958) The distribution of freshwater invertebrates in ponds in relation to annual fluctuations in the chemical composition of the water. J. Anim. Ecol., V.27 : pp. 105-23.

Wadajo, K. and Belay, A., (1984) Species composition and seasonal abundance of zooplankton in two Ethiopian Rift Valley Lakes. Hydrobiologia, V. 113: pp. 129-36.

Welch, P.A., (1948) Limnological methods. Bleksiton, Philadelphia, 381p.

Winberg, G.G., (1956) Rate of metabolism and food requirements of fishes. Nauchnye Trudy Belorusskova Gosudarstvenoyo Universiteta imeni, V.I. Leniana, Minsk. 253 p.

Zafar, A.R., (1964a) On the ecology of algae in certain fish ponds of Hyderabad India. I. Physico-chemical complexes. Hydrobiologia, V. 23: pp. 179-95.

Zafar, A.R., (1967) On the ecology of algae in certain fish ponds of Hyderabad, India, III. The periodicity. Hydrobiologia, V. 30: pp. 96-112.

12

Impact of Groundwater Development and Industrial Pollution of Environment

N KITTU

AND

P. R. SUBRAMANIAN

Introduction

Fresh water constitutes a mere 2.67% of the total global water resources, of 1,460 million cubic kilometres. Of these reserves a major part is locked up in the form of polar ice and glaciers. Groundwater accounts for about 23% of all fresh water resources available and hence its proper development and judicious exploitation assumed a great importance. The most up-to-date estimate by the Central Ground Water Board put the annual exploitable ground water potential at 42.3 million hectare metres (mham), of which only 10 mham are being exploited at present. Recent investigations show that the hard rock formations of Peninsular India - which constitute nearly 70 per cent of the country's landmass hold more groundwater than was assumed earlier.

Groundwater development progressed steadily through the five year plan periods. From a mere 6.5 million hectares in 1950-51 it came to provide 35 million hectares of irrigation by the end of VII plan period. The importance accorded to the groundwater sector would be apparent from the fact that it accounts for 55% of the total additional irrigation facility. This massive development of groundwater coupled with the introduction of the high yielding varieties of seeds was largely responsible for the "green revolution" witnessed in our country during the sixties, making us self sufficient in the production of food grains (Central Ground Water Board, 1989).

Tube well irrigation in India, which began about 50 years ago has seen rapid development in the last two decades on account of the nationwide droughts and bringing greater area under cultivation. However, development of the groundwater resources in an unplanned manner without any consideration to the safe yield, safe spacing between wells, etc., have resulted in over development of the aquifer. Discharge of untreated industrial effluents have resulted in large scale pollution.

Effect of mining operations have had deleterious effects on the groundwater regime. These are discussed with a few case histories.

I. Water level decline

The spurt in the number of tube wells is an indication of the increasing use of groundwater. Groundwater withdrawals consistently exceeding the recharge lead to a permanent decline in the water table.

(*i*) *Maharashtra* : In Maharashtra, withdrawals of a groundwater in excess of the annual recharge for cultivation of water intensive crops has led to serious drinking water and food crisis. Between 1960-61 and 1980, the number of wells in the state was increased by 51 per cent but the area irrigated by them was doubled. This was mainly due to the spread of the use of mechanised pump sets. The rapid depletion of groundwater has increased the number of "Problem Villages" with no source of drinking water from about 17,000 in 1980 to 23,000 in 1983.

The shift in cultivation from rainfed coarse grains to the water intensive sugarcane crop in Maharashtra has also led to severe local shortage of groundwater.

(*ii*) *Gujarat* : Much the same thing is happening elsewhere. The Ahmedabad Municipal Corporation increased its groundwater withdrawals from 1.7 billion gallons (in 1951-52) to 18.32 billion gallons (in 1971-72). It was reported that in 1980, 450 tubewells withdrew 34 billion gallons to meet domestic and industrial demand. It was also reported that in the recent past, the level has been going down at a rate of 1.5 to 1.9 metres per year. But lately, a drop of 3 metres to 6 metres in a season has also been noticed in some areas like Gomtipur. Consumers who do not have access to pumps with higher lifts face increasing hardship. The Sabarmati river bed is dry for most of the year, partly because groundwater seepage into the river is almost negligible.

II. Saline water incursion

Over development of groundwater can lead to the intrusion of saline water, particularly in coastal areas, making the water unfit for drinking and irrigation. In the fifties, farmers in the southern coastal plains of Saurashtra in Gujarat, began intensive cultivation of vegetables and sweet lime using groundwater. By 1970, the groundwater was affected by saline water intrusion from the sea and a drop in agricultural production followed. Between 1971 and 1977 the affected area increased from 35,000 hectares to 1,00,000 hectares.

The deltas of the Godavari, Krishna and Cauvery rivers on India's east coast represent important pockets of groundwater. In the past, water was drawn from shallow wells and there were return flows and proper drainage control. But in the last two decades, exploitation of these reservoirs has mushroomed with the increased emphasis on multiple cropping and high yielding varieties. The recurring droughts in Tamil Nadu have led to the installation of a large number of deep bore wells along the coast. This has hightened the threat of saline intrusion, particularly in the tiny coastal groundwater reservoir around Madras city.

III. Industrial Pollution

Groundwater is less susceptible to pollution than surface water, but its pollution is more difficult to detect or rectify. Recent studies indicate that the problem is growing seriously. Some of the culprits are textile-printing and dyeing units in Rajasthan, the coir processing industry in Kerala, tanneries in Tamil Nadu and chemical industry in Karnataka.

1. *Textile dyeing units, Jodhpur* : In September, 1983, several newly installed tube wells and hand pumps near an industrial area near Jodhpur discharged coloured water. A report by the Jodhpur Environment Cell of the Gandhi Peace Foundation (GPF) reveals that about 1,500 small scale textile units in Jodhpur, Pali and Balotra discharge 15 million litres of effluents every day into open drains, river beds, reservoirs and ditches. Because the soil is sandy, the harmful chemicals percolate into wells, tanks and reservoirs from which about a million people obtain their water. Cattle, wildlife and crops are also affected (Centre for Science and Environment, 1984-85). A GPF survey identified various forms of cancer, among other diseases, in the area. The high pH of the water-more than 9, has affected crops. The GPF report alleges that 7,000 hectares to 10,000 hectares of land has been ruined and charges that the industry which has an annual turnover of Rs. 400 crore is taking no pollution control measures.

2. *Coir Industry, Kerala* : The Coir processing industry in Kerala is destroying water in much the same way. All across the state, coir units soak coconut husks in pits to remove the fleshy matrix from the fibre. In the process, groundwater is contaminated by hydrogen sulphide and organic acids.

The shallow aquifers could be flushed by heavy rain but the problem of the deeper aquifers is more acute as they are the main source of water for the state in the future.

3. *Tanneries in Tamil Nadu* : The North Arcot district of Tamil Nadu is best example of how industry can ruin the environment. Untreated effluents from about 250 tanneries most of which abound in the towns of Ambur and Ranipet, have turned it into a wasteland. Crop production is down by more than 70 per cent. The water is undrinkable. The level of water conductivity, which determines the "safeness" of water, is several hundred times the permissible limit. Respiratory and skin diseases have a field day. Even animals refuse to graze in these stinking parts (Centre for Science and Environment, 1984-85).

Nearly 70 per cent of India's total exports of leather and leather products come from the 423 recognised tanneries in Tamil Nadu. There is a concentrated attack on both soil and water in Ranipet, Ambur and Vaniyambadi in the district. Of these Ambur is the worst affected. The daily discharge of effluents from 41 tanneries in Vaniyambadi, is around 1,100 cum. Many tanneries have installed concrete storage tanks from which the effluents are evaporated. But their capacity is limited and the overflow is dumped either on fields or let into irrigation tanks and rivers. The King's Institute in Madras has declared Ambur's water supply to be unfit for human consumption.

The effluents from the tannery industry have resulted in 10,000 acres being severely affected and another 30,000 acres being moderately affected in Vaniyambadi, Ambur, Vellore, Arcot, Wallajah, Ranipet, Visharam and Timir towns. Soil in these areas yields only about 20 tonnes of sugarcane an acre as against the normal yield of 60 tonnes. In Pernambut, 3,000 acres of prime land are being used for ragi which is believed to better withstand the effluent onslaught, but the ragi output is also falling.

What compounds the problem is that the effluents are released directly into or find their way to the Palar river and all wells on both banks and the river itself the major source of water supply-have been severely polluted. Some 3,000 taps in Ambur now disgorge only polluted water.

4. *Chemical Industry, Karnataka* : The Harihar Polyfibres Factory in Harihar town, Dharwad, produces about 165 tonnes of factory grade pulp every day, which is used by the nearby plant of the Gwalior Rayon and Silk Manufacturing Company (GRASIM) to make rayon. Both plants also produce 45,000 cum of effluents daily. Despite claims that the discharge is being properly treated before it is let loose into the Tungabhadra river, the effluents are clearly a menace to the environment and to people (Centre for Science and Environment, 1984-85).

The once sparkling water of the Tungabhadra, is now stinking, brownish red. Frothy black liquid and pale green alkali wash travels down to the river under a smelly canopy of sulphide fumes. Gases like sulphur dioxide, carbon monoxide and hydrogen sulphide play havoc with the lungs of the villagers.

The results are obvious. Almost all the villages downstream are forced to use the polluted water and complain of frequent headaches and leg and back pains. Dysentery, indigestion and gas are common as are skin irritation and erruptions. The eyes turn yellow. There is also an unknown malady of the intestines which calls for costly surgical treatment.

Most villagers are dependent on the river for drinking water, bathing, irrigation, fishing and for giving water to their animals. With the river polluted, the daily collection of drinking water has become a slow, torturous procedure.

Crops are drastically affected. It is said that where one used to get 6 to 7 tonnes of Jowar from one hectare, one now gets just 1.8 to 2 tonnes. Groundnut yields have dropped by 10 to 15 bags. One prosperous farmer says he knew the factory was the culprit because the yield went up when the factory was shut down for six months due to a strike.

A factory report claims that the water is polluted by the Belludi Nala and Harihar towns sewer discharge even before the effluents get into the river. However, the Karnataka University, Dharwad, analysed samples of effluent and river water and reported that the levels of BOD and COD after the affluents meet the river go up as high as 300 and 1,200 respectively. The factory administration says that it is improving its treatment from water of irrigation standards to water of river discharge standards. Yet there are regular incidents of mass mortality of fish.

In February, 1984 there was a massive fish kill. The villagers did not take this lying down and samples of the dead fish reached the relevant authorities while letters went to the Chief Minister. The State Pollution Control Board ordered the factory to stop the effluent discharge and its special team of investigators found dissolved oxygen levels were so low that the fish could not live. Only when the dissolved oxygen content rose to acceptable levels was the ban lifted.

IV. Effects of Mining Activities

Iron ore mining in Goa : Iron ore mining is an important economic activity in Goa. Reserves of iron ore are estimated to be 480 million tonnes. Annual production is 13.8 million tonnes. There are 76 iron ore mines in Goa. Mining is largely mechanised and transportation is through major water ways (Venkatraman, 1984).

Environmental impact of iron ore mining operation : The large scale mining operations in Goa district have had adverse effect in many aspects of environment, agriculture, water resources, forest cover, etc. The following are some of the important ones.

1. Pumping of highly turbid and muddy storm water has led to siltation of water ways resulting in more areas becoming flood-prone.
2. Stagnant water in the abandoned pits is a health hazard.
3. Large scale stripping of laterite cover in the mining areas affects the area and quantum of groundwater recharge to laterite aquifers.
4. Denudation of forests and removal of vegetal cover on the hill faces increase run-off and flood hazard and reduces infiltration of rainfall.
5. Candepar, Bicholim, Madei (Usgaon) are some of the major rivers whose waters have been polluted by the mine wastes and washed out materials from the dumps; some of which are located very close to the river. Candepar river is a source of water supply to Goa.
6. Mine wastes in the form of huge dumps (both active and dead) occupy substantial areas. These form the perennial source of silts which in turn are washed into water ways during heavy monsoon rainfall.
7. Damage to agricultural land is considerable. It is reported that 253 ha. of prime agricultural land have been adversely affected.
8. Turbid waters and suspended matter in water courses affect aquatic life.
9. Atmosphere in the mining areas is dustladen which has settled down as a thick layer on buildings, trees, roads and these are picked up again and again by the passing heavy earth moving machinery and transport trucks.
10. Large scale mining operations involving heavy blasting may have adverse long term effects on fauna and flora.

Remedial Measures

1. To avoid indiscriminate sinking of new wells, especially in over developed areas indicated by steady decline of water level in the wells, control should be exercised before advancing loan facilities for well sinking, allowing electrical connections to pump sets, etc.
2. The norms recommended for safe spacing between wells as also the safe pumping limits from individual wells must be enforced.

3. Preliminary surveys must be taken up in areas proposed to be brought under canal irrigation to identify factors like the optimum input of water, the soil profile, type of cropping pattern to be practised, etc. It has been noticed that in many canal command areas the farmers whose lands are located nearer the major distributaries tend to apply excessive water, resulting in water-logging. Educating the farmers about the crop water requirement and modern methods of agriculture like mixed cropping, plantation of eucalyptus trees, etc, would help in reducing the water-logging problems. An effective system of drains also must be constructed along with the network canals.
4. Control of groundwater extraction through pumping from the coastal aquifers by enforcing optimum rate of withdrawal, safe spacing, etc., the landward migration of fresh/saline water interface can be checked.
5. To lay down norms for effluent discharge from factories and ensure their strict implementation. Make the industries adhere to these norms and make them liable for deleterious effects on the environment.

References

Central Ground Water Board, (1989) Bhujal News, V. 4, No. 1.

Centre for Science and Environment, (1984-85). The state of India's Environment II, the Citizen's Report.

Venkatraman, R., (1984) Preliminary report of studies on the effects of iron ore mining operations on groundwater resource in Pale Surla mining areas, Goa.

Journal of CGWB., (1989) Bhujal News, V. 4, No. 1. •

13

A Study on the Mass Removal of Unwanted Fish in the Carp Culture Ponds of Lower Deltaic Region of River Krishna, South India

P. MADHAVEELATHA

AND

M.K. DURGA PRASAD*

Introduction

The lower deltaic region of river Krishna, South India supports diversified aquatic habitats (rivers, canals, lakes, ponds, etc.). A variety of fish species are inhabiting these water bodies. Till 1970, people of this region depend on fish captured from these natural water bodies. However, with the increased demand of fish, since 1974 several agriculturists of this region converted their low lying lands (latitudes 16°10′ and 15°20′N and longitudes 80°20′ and 80°E) into a number of carp culture ponds. These ponds vary in size from 0.5 hectare to 20 hectares and the total area of these ponds is estimated to be 100,000 hectares. In all these water bodies composite carp culture is practised. The Indian major carps, viz., *Catla catla, Labeo rohita and Cirrhinus mrigala* are commonly used for semi intensive culture. In addition to them, rarely, two exotic fish, namely, *Cyprinus carpio* and *Ctenopharyngodon idella* are also introduced by some farmers. These fish under better management conditions exploit different niches in the carp culture ponds and yield a good production (16.28/ha). However, one of the major problems posed for culturing these carps is the competition between desirable carp species for food and space with the other fish species trespassing into the culture ponds. These undesirable fish species enter the fish ponds either along with the carp seed or through the inlet waters of the pond in the form of eggs, juveniles or adults. In the present study, an attempt is made to identify these "unwanted fish" species. Further, a study has been made on the mass removal of these species by employing different methods. The importance of the control of these unwanted fish was emphasised by Bennet (1962), Jhingran (1965 & 1968) and Natarajan, *et al.* (1975).

Material and Methods

To identify the unwanted fish an extensive survey was made in the carp culture ponds of the lower deltaic region of Krishna, South India. Unwanted fish species were collected twice in every month from May, 1988-January, 1991 by using cast net with 1.25 cm. size mesh. These species were preserved in 10% formalin. For identification of these fish, Day (1875 to 1878), Hubbs and Lagler (1947) and Jayaram (1981) were consulted. The gut contents of all these fish were analysed to understand the type of food they are ingesting and the food items were expressed in percentages following the method suggested by Hunes (1950) and Pillay (1952). Three ponds were chosen to study the impact of unwanted fish on carp production and the methods employed for their mass removal in the carp culture ponds. A culture pond located at Sreeparru (4 ha) was designated as Pond 'A'. In this carp culture pond unwanted fish were not removed and the fish were harvested after 8 months. An another pond located at Budham (1.2 ha) was designated as Pond 'B'. In the carp culture pond unwanted fish were partially removed by employing repeated netting method (2.5 cm. mesh size cast net). An another carp culture pond located at Potumarru (1.6 ha.) was designated as Pond 'C'. In this pond unwanted fish were successfully eliminated by 1. installing 'O' size mesh at the inlets of culture pond (double filtration method), 2. treating the pond with organophosphate piscicide, Ekalux (quinalphos) 160 ml/ha/m., 3. repeated netting by using 2.5 cm. mesh size cast net and 4. introducing predatory fish *Wallago attu* (15 numbers/ha). In all these carp culture ponds, trial nettings were made to capture the unwanted fish at every month, during the culture period. The number of nettings made at every month, the total number of fish captured, the average number of fish per one netting and the total weight of fish expressed in kilograms on the day of investigation were calculated. The order of dominance of different species in the corresponding nettings was also noted. Since the duration of the culture period is varying in the three ponds, A, B and C, the total weight of fish harvested is computed for 365 days and total weight of the fish is expressed as tonnes/ha/365 days. The cost of the fish harvested in these ponds is also computed for 365 days.

Results

In the present study 44 unwanted fish species were identified from different carp culture ponds located in the lower deltaic region of river Krishna, South India (Table 13.1). Among them, only 15 species are common to majority of ponds investigated. Further it is noted that these

Table 13.1 : List of Unwanted Species Encountered in Carp Culture Ponds

S. No	*Name*
1.	*Elops machnata* (Forkal).
2.	*Megalops cyprinoides* (Broussoner).
3.	*Anguilla bengalensis* (Gray & Hardwy).
4.	*Hilsa ilisha* (Ham).
5.	*Setipinna breviceps* (Cantor).
6.	*Notopterus notopetrus* (Ham).
7.	*Chanos chanos* (Forskul).
8.	*Chela labuca* (Ham).
9.	*Salmostoma bacaila* (Ham).
10.	*Esomus danricus* (Ham).
11.	*Rasbora daniconius daniconius* (Ham).
12.	*Amblypharyngodon mola* (Ham).
13.	*Puntius sophore* (Ham).
14.	*Puntius chola* (Ham).
15.	*Puntius sarana* (Ham).
16.	*Osteobrama cotio cotio* (Ham).
17.	*Labeo calbasu* (Ham).
18.	*Cirrhinus reba* (Ham).
19.	*Lepidocephalichthys thermalis* (Cuv. & Val.).
20.	*Mystus cavasius* (Ham).
21.	*Mystus Vittatus* (Bloch).
22.	*Mystus bleekeri* (Day).
23.	*Ompok bimaculatus* (Bloch).
24.	*Wallago attu* (Bloch).
25.	*Clarias batrachus* (Linnaeus).

44 species are distributed in different culture ponds in different combinations. Ponds nearer to the coast are often invaded with estuarine fish, namely, *Megalops cyprinoides, Lates calcarifer* and *Setipinna breviceps, Hilsa ilisha,* an anadromous fish is also noticed in one of the ponds investigated. Marine fish species, namely, *Scatophagus arus, Chanos Chanos, Elops machanta* appeared in the fish ponds of Khajipalem located nearer to Bay of Bengal. The ponds located in and around lake Kolleru (16°44′ N and 81°72′E) are dominated by an undesirable fish *Tilapia mossambica.* The other fish species are more or less common to all the ponds investigated. However, in the present study, it is observed that only 15 species of fish appear in relatively good numbers in several ponds. All these fish feed on a variety of food items (Table 13.2). Depending on the type of food taken by the unwanted fish, it is possible to group the unwanted fishses as : 1. Planktonivorous, 2. Herbivorous, 3. Carnivorous and 4. Carnivorous and predatory fish.

An attempt has been made to study the abundance of unwanted fish in relation to the different mass removal methods adopted in different ponds. In pond 'A' the Indian major carps were cultured without removing the unwanted fish. In this pond, trial nettings were performed at every month (except August) during the culture period. The number of trial nettings performed, the total number of fish removed in trial nettings, the average number of fish per one netting, the total weight of fish removed through trial nettings and the order of dominance of unwanted fish species at different months during culture period are presented in Table 13.3. Further the total weight of the unwanted fish removed in trial nettings and at the time of harvest, the total weight of the carps harvested and the total weight of fish (Carps + unwanted fish) harvested are presented in Table 13.3. The approximate cost of these fish is also presented.

In pond 'B' the unwanted fish entered the pond and colonised along with the carps. They are removed only by the repeated netting method. In this pond, trial nettings were performed at different months during the culture period. The number of trial nettings performed, the total number of fish removed in trial nettings, the average number of fish per one netting, the total weight of fish removed through trial nettings and the order of dominance of unwanted fish species at different months during culture period are presented in Table 13.4. Further, the total weight of the unwanted fish removed in trial nettings and at the time of harvest, the total weight of the carps harvested and the total weight of fish (Carps + unwanted fish) harvested are presented in Table 13.4. The approximate cost of these fish is also presented in the same Table.

Table 13.2 : Percentage Composition of Food Items in Commonly Occurring Species of Unwanted Fish in Carp Culture Ponds of Lower Deltaic Region of River Krishna

	Algae	*Rotifers*	*Worms*	*Crutaceans*	*Insects*	*Fish*	*Fish Scales*	*Sand Mud*	*Supplement-ary Feed*	*Name of the Fish*
1	**2**	**3**	**4**	**5**	**6**	**7**	**8**	**9**	**10**	**11**
	90	–	–	–	–	–	–	–	10	*A. mola*
Herbivo-	40	4	–	1	–	5	2	–	48	*T. Mossambica*
rous.	70	23	–	–	–	–	–	–	7	*E. danricus*
	15	20	–	50	–	–	–	–	15	*R. maculatus*
	35	8	–	25	–	–	–	–	31	*P. sophore*
Planktonivorus	6	26	–	27	–	–	2	–	40	*P. chola*
	1	5	–	50	–	–	–	–	42	*S. bacaila*
	–	27	–	40	15	–	–	–	20	*C. nama*
	–	5	–	10	30	3	5	–	45	*G. giuris*
Carnivo-	–	8	–	3	25	–	15	4	50	*M. vittatus*
rous.	5	–	20	15	27	–	3	–	30	*C. batrachus*
	5	1	–	5	15	28	–	10	36	*C. punctata*
Carnivorous	10	–	40	–	15	10	–	6	19	*C. punctata*
&	10	–	40	–	15	10	–	6	19	*N. notopterus*
Predatory	2	1	10	–	10	67	–	4	5	*W. attu*
	25	–	–	–	–	–	18	17	40	*M. pancalus*

Table 13.3 : Fish Culture Pond-A, Sreeparru Carps Cultured without Removal of Unwanted Fish

CONDITIONS :

1. Extent of pond : 4 hectares
2. Date of introducing the carp seed : 13-6-89
3. Weight of the seed fish stocked : 100 to 150 gm.
4. No. of seed fish stocked : 3,000/hectare
5. Fertilisation of the pond : Improper fertilisation
6. Quantum of supplementary food : 4% of the body weight at 4 day interval
7. Date of harvest : 10-3-90
8. Methods employed for removal of unwanted fish : Nil
9. Approximate expenditure incurred for elimination of unwanted fish : Nil

S. No.	*Date of trial netting*	*No. of trial nettings performed*	*Total no. of fish removed in trial netting*	*Average no. of fish per one (Kg)*	*Total wt. of unwanted fish removed*	*Order of dominance of unwanted fish species*
1.	6-07-'89	8	17,184	2,148	71.736	*T. mosambica* > *E. danricus* > *P. sophore* > *M. vittatus* > *A. testudineus*
2.	8-09-'89	7	32,054	44,579	160.27	*T. mosambica* > *E. danricus* > *P. sophore* > *M. vittatus* > *A. testudineus*
3.	21-10-'89	11	12,680	1,152	150.02	*T. mosambica* > *E. danricus* > *S. bacaila* > *P. sophore* > *M. vittatis*
4.	18-11-'89	5	8,162	1,632	40.81	*T. mosambica* > *E. danricus* > *P. sophore* > *M. vittatus* > *A. testudineus*
5.	14-12-'89	9	19,235	2,137	76.94	*T. mosambica* > *E. danricus* > *P. sophore* > *S. bacaila* > *M. vittatus*
6.	26-12-'89	4	21,133	5,283	84.53	*T. mosambica* > *E. danricus* > *S. bacaila* > *P. sophore* > *M. vittatus*
7.	4-01-'90	9	29,166	3,240	98.60	*T. mosambica* > *E. danricus* > *S. bacaila* > *P. sophore* > *M. vittatus*
8.	24-01-'90	11	12,330	1,121	49.32	*T. mosambica* > *E. danricus* > *S. bacaila* > *P. sophore* > *M. vittatus*
9.	12-02-'90	6	19,855	3,309	92.75	*T. mosambica* > *E. danricus* > *S. bacaila* > *P. sophore* > *M. vittatus*

(Contd.)

Table 13.3 (Contd..)

Fish harvested in Pond-A (tonnes/4.0 hectares)		*Wt. of fish for a culture period (tonnes/hectare)*	*Cost of Rupees (US$)*	*Wt. of fish computed/hectare/*	*Cost in Rupees (US$) 365 days*
Wt. of unwanted fish :					
(*a*) Removed in trial nettings	0.825				
(*b*) Removed at the time of harvest	5.00	1.456	4,368 ($145.6)	2.15	6,045 ($201.3)
Total Wt. of unwanted fish	5.825				
Total Wt. of the carps harvested	7.50	1.875 ($1375)	41.250	2.77	60,940 ($2031.3)
Total Wt. fish	13.325	3,331	45,618 ($1520.6)	4.92	66,985 ($2232.8)

Table 13.4 : Fish Culture Pond-B, Buddham Carps Cultured with Partial Removal of Unwanted Fish

CONDITIONS :

1. Extent of pond : 1.2 hectares
2. Date of introducing the carp seed : 2-6-89
3. Weight of the seed fish stocked : 100 to 250 gm.
4. No. of seed fish stocked : 1,840/hectare
5. Fertilisation of the pond : Both organic and inorganic were used
6. Quantum of supplementary food : 5% of the body weight of the fish
7. Date of harvest : 26-11-89
8. Methods employed for removal of unwanted fish
   (*a*) Repeated nettings
9. Approximate expenditure incurred for elimination of unwanted fish : Rs. 1000/hectare prior to and after stocking the fish seed

S. No.	*Date of trial netting*	*No. of trial nettings performed*	*Total no. of fish removed in trial netting*	*Average no. of fish per one (Kg.)*	*Total wt. of unwanted fish removed*	*Order of dominance of unwanted fish species*
1.	27-07-'89	6	1,760	293	35.20	*P. chola* > *P. sophore* > *C. nama* > *E. danricus*
2.	13-08-'89	4	1,840	460	36.80	*P. chola* > *P. sophore* > *C. nama* > *E. danricus*
3.	20-09-'89	9	2,512	279	75.96	*P. chola* > *P. sophore* > *C. nama* > *E. danricus*
4.	21-09-'89	13	12,000	923	300.00	*P. chola* > *P. sophore* > *C. nama* > *E. danricus*
5.	22-09-'89	9	3,560	395	106.80	*P. chola* > *P. sophore* > *C. nama* > *E. danricus* > *M. cyprinodes*
6.	13-10-'89	10	5,687	569	170.61	*P. chola* > *P. sophore* > *C. nama* > *E. danricus*
7.	24-11-'89	9	3,814	424	152.56	*P. chola* > *P. sophore* > *C. nama* > *E. danricus* > *M. cyprinoides*
8.	25-11-'89	8	2,896	362	114.04	*P. chola* > *P. sophore* > *C. nama* > *E. danricus*

Table 13.4 (Contd.)

Fish harvested in Pond-A (tonnes/4.0 hectares)		*Wt. of fish for a culture period (tonnes/hectare)*	*Cost of Rupees (US$)*	*Wt. of fish computed/hectare/ 365 days*	*Cost in Rupees (US$)*
Wt. of unwanted fish :					
(*a*) Removed in trial nettings	0.991				
(*b*) Removed at the time of harvest	0.6	1.325	3,975	3.28	9,840
Total Wt. of unwanted fish	1.591				
Total Wt. of the carps harvested	3.75	3.12	66,264	7.746	1,70,412
Total Wt. fish	5.341	4.445	70,239	11.026	1,80,252

In Pond 'C' the unwanted fish were eliminated by employing different methods at various stages of culture. In this pond also trial nettings were performed at different months (except October) during the culture period, the number of trial nettings performed, the number of fish removed in trial nettings, the average number of fish per one netting, the total weight of fish removed through trial nettings and the order of dominance of unwanted fish species at different months during culture period are presented in Table 13.5. Further the total weight of unwanted fish removed in trial nettings and at the time of harvest, the total weight of the carps harvested and the total weight of the fish (carps + unwanted fish) harvested are presented in Table 13.5. The approximate cost of these fish is also presented in the same Table.

Discussion

In the present study, a total of 44 unwanted fish species are recorded. Though these fish species differ in their feeding habits, compete with the major carps. Among these, the commonly occurring 15 species severely compete with major carps for natural food and supplementary food, (Table 13.2). Due to this competition the growth rate of carps is drastically effected. A study on the production of fish in ponds. A, B & C can further substantiate this fact. In Pond (A) carps were harvested after 247 days whereas in Pond (B) they were harvested only after 147 days. In Pond (C) the harvesting was taken up after 328 days. Hence, for the sake of comparison the weight of the unwanted fish, carps and the total weight of the fish in all these ponds were computed for 365 days and the data is presented in the corresponding tables. These figures indicate that in Pond 'A' both unwanted fish and carps were grown more or less to the same magnitude and the total production of fish computed per 365 days would be 4.92 tonnes and the cost of this fish is Rs. 66,985/. Though the fish production is significant, due to low cost of unwanted fish (Rs. 3/ Kg.) the total amount realised from the harvest is considerably low.

In Pond 'B' though the unwanted fish are removed by repeated netting, these fish were not totally eliminated. As such, the total weight of unwanted fish harvested is 3.28 tonne/ha/365 days. In this Pond, the carps harvested are estimated to be 7.76 tonnes/ha/365 days. When a comparison is made between Pond A & B it is clear that in Pond 'B' the total yield of fish is high. This is due to the fact that the stocking size of fish in Pond 'B' is higher than the stocking size of fish in Pond A. Further, the supplementary feed is given at the rate of 5% to the body weight of fish per day in Pond 'B'. Whereas in the Pond 'A' the fish are

Table 13.5 : Fish Culture Pond-C, Potumarru Carps Cultured by Employing Four Different Methods for The Removal of Unwanted Fish

CONDITIONS :

1. Extent of pond : 1.6 hectares
2. Date of introducing the carp seed : 25-5-89
3. Weight of the seed fish stocked : 50 to 100 gm.
4. No. of seed fish stocked : 8,750/hectare
5. Fertilisation of the pond : Both organic and inorganic fertilisers were used prior to and after stocking the fish seed
6. Quantum of supplementary food : 6% of the body weight of the fish
7. Date of harvest : 17-4-90
8. Methods employed for removal of unwanted fish :
   (*a*) Double filtration method used to filter the unwanted fish at the inlet by installing 0-mesh size nets
   (*b*) The culture pond is treated with organophosphate piscicide, Ekalux (Quinolphos), at the rate of 15 numbers before stocking the fish
   (*c*) Predatory fish *Wallago attu* stocked at the rate of 15 number per hectare after one month of stocking of carp seed
   (*d*) Repeated netting performed to remove the unwanted fish
9. Approximate expenditure incurred for elimination of unwanted fish : Rs. 2000 (US$ 66.6) hectare

S. No.	*Date of trial netting*	*No. of trial nettings performed*	*Total no. of fish removed in trial netting*	*Average no. of fish per one (Kg.)*	*Total wt. of unwanted fish removed*	*Order of dominance of unwanted fish species*
1.	16-05-'89	10	1,118	112	22.36	*P. sophore > S. bacaila > M. pancalus > M. vittatus*
2.	1-06-'89	6	896	149	22.40	*P. sophore > S. bacaila > M. vittatus > C. batrachus > M. pancalus*
3.	12-06-'89	5	672	134	13.44	*P. sophore > M. vittatus > S. bacaila > A. testudineus*

4.	15-07-'89	4	189	47	9.45	*P. sophore > S. bacaila > M. vittatus*
5.	15-08-'89	8	560	70	14.10	*S. bacaila > M. vittatus > P. sophore*
6.	27-09-'89	7	570	81	14.45	*S. bacaila > A. testudineus > M. vittatus*
7.	27-11-'89	4	115	29	5.75	*S. bacaila > P. saophore > M. vittatus > M. pancalus*
8.	31-12-'89	3	96	32	3.50	*S. bacaila > M. vittatus > C. cyprionoides*
9.	27-1-'90	3	18	6	0.072	*S. bacaila > M. armatus*
10.	27-2-'90	3	6	2	0.024	*S. bacaila > M. armatus*
11.	21-3-'90	2	4	2	0.005	*S. bacaila > M. armatus*

Fish harvested in Pond-C	*(tonnes/4.0 hectares)*	*Wt. of fish for a culture period of 328 days (tonnes/hectare)*	*Cost of Rupees*	*Wt. of fish computed/hectare/ 365 days*	*Cost in Rupees*
Wt. of unwanted fish :					
(*a*) Removed in trial nettings	0.105				
(*b*) Removed at the time of harvest	0.075	0.1125	337	0.125	375
Total Wt. of unwanted fish	0.180				
Total Wt. of the carps harvested	23,408 ($10728.6)	14.63	321,860	16,280	358,160
Total Wt. fish	23,588	14.74	325,235	16,405	358,535

fed with supplementary feed at the rate of 4% to the body weight at four day interval even though the stocking density in this Pond is much higher than that of Pond 'B'. Hence the total production of the fish in Pond 'B' is high compared to Pond 'A'. However, due to lack of proper methods in eliminating the unwanted fish, these fish populations in Pond 'A' grow enormously by making use of the optimal conditions. Hence their production is relatively high to that of the cultured carps. Further, the repeated netting method employed is not successful in eliminating these unwanted fish in Pond 'B'. Though the total fish production is high in this pond (11.026 tonnes/ha/365 days), the amount realised by the farmer is only Rs. 1,80,252/ha/365 days. This is due to the fact that the unwanted fish fetched a low price to the farmer (Rs. 3/ per Kg.).

In Pond 'C', a sincere attempt is made to remove the unwanted fish by employing four different methods (Table 13.5). Further, the Pond was fertilized regularly and supplementary feed is given at the rate of 6% to the body weight of the fish. Due to successful elimination of unwanted fish, the carps have fully utilized the ecological resources and gave an yield of 16.28 tonnes/ha/365 days. Whereas the unwanted fish were harvested to a tune of 0.125 tonnes/ha/365 days. The total weight of the fish (cultured carps + unwanted fish) realised from this pond is 16.405 tonnes/hectare/365 days and this has fetched Rs. 3,58,535.

From this it can be concluded that in a successful fish farming, the mass removal of unwanted fish plays a considerable role and the combination of four methods employed for the mass removal of fish in pond 'C' is highly useful for the farmers for achieving better yield of fish. Further, the cost of expenditure incurred for elimination of unwanted fish by using these four methods is only Rs. 2,000/-/hectare.

References

Benett, G.W. 1962, Management of artificial lakes and ponds, Rinhold Publishing Corporation, New York.

Day, F, 1875-1878, "The Fishes of India" being a natural history of the fishes known to inhabit the Seas and fresh waters of India, Burma and Ceylon. Text and Atlas in 4 Parts. London xxx 778 pp; 195 Pls.

Hubbs, C.L. & Lagler, K.F. 1947, "Fishes of great lakes region", Bull. No. 26, Cranbrook Inst. Sci., 196 pp.

Hynes, H.B.N. 1950, "The food of freshwater Sticklebracks (Gastrosters aculeatus and Pyrosteus Pongitius) with a review of Methods used in studies of the food of Fishes", J. Anim Ecol. Vol. : 19. pp. 36-58.

Jayaram, K.C. 1981, "The freshwater fishes of India, Pakistan, Bangladesh, Burma & Sri Lanka", Publi. Zool. Surv. India, Calcutta, 475 pp.

Jhingran, V.G. 1965, "Report on inland Fisheries research and management and fish culture in U.S.S.R." Cent. Inl. Fish Res. Inst. Mics. Contr. (5) Mimco.

Jhingran, V.G. 1983, Fish & Fisheries of India, Hindustan Publication, New Delhi, India, 1983 Edition, pp. 369-71.

Natarajan, A. V. 1971, "Problems of Fisheries development in reservoirs", Proc. First workshop on All India Co-ord. Res. Proj. on Ecology and Fisheries of Freshwater reservoirs. Central Inland Fisheries Research Institute, Barrackpore, 30th, 31st Aug., 1971.

Natarajan A.V., M. Rama Krishnaiah & M.A. Khan, 1975, "The food spectrum of trash fish in relation to major Carps in Konar & Tilaiya Reservoirs (Bihar)", J. Inland Fish. Soc. India, Vol. VII, pp. 65-75.

Distribution, Composition and Numerical Abundance of Benthic Macrofauna of a Tropical Lake

S. M. ANVAR BATCHA

Introduction

The importance of macrobenthic fauna has been well understood in recent years as an essential part of the trophic cycle and as indicator organisms in discussing the conditions of nature, and characteristic of that niche (Miyadi, 1941a and Kitamori, 1963). They are also sensitive to environment and are used by water quality biologists to detect effects of environmental perturbations. The macrobenthic fauna of Vembanad Lake has been studied by various authors in relation to its distribution, composition and abundance (Desai and Krishnmkutty 1967a, Kurian, 1972, Kurian *et al.*, 1975, Anzari, 1977 and Pillai 1977 and 1978). All these investigations were carried out before the construction and commissioning of the Thanneemukkams salt water barrier across the Vembanad Lake.

Vembanad Lake-Area of Study

Vembanad Lake is the largest backwater system found along the South-West Coast of India. It extends between latitudes 9°28′, and 10°10′N and longitude 76°13′ and 70°30′E. Lengthwise it starts at Azhikode in the north and ends in Alleppey in the south with about 90 km. distance and a total area of about 300 sq. km (Fig. 14.1). Of the six major rivers which discharge fresh water into the lake, two of them, namely, Periyar and Muattupuzha open into the northern part of the Thanneermukkam bund, whereas the other four, namely, Achankoil, Pamba, Manimala and Minachil open into the southern part of the barrier. The fresh water flowing from the southern part of the lake to the northern part is virtually blocked by the Thanneermukkam barrier (Batcha and Damodaran, 1987).

Material and Methods

The materials for the present study were collected from 17 stations which are situated at equal distance, starting from Cochin barmouth to Vaikom Vetty (Fig. 14.1). The quantitative macrofaunal samples were

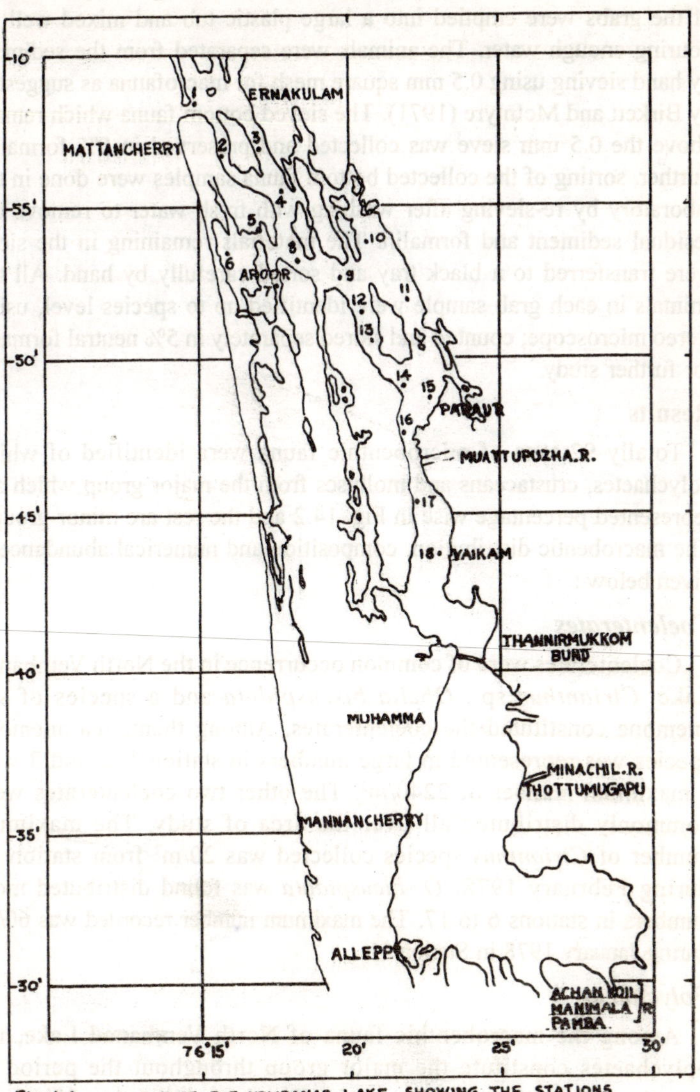

FIG.14·1 MAP OF THE VEMBANAD LAKE SHOWING THE STATIONS

collected at monthly intervals using 0.1 m^2 van Veen grab. The contents of the grabs were emptied into a large plastic tub and mixed well by pouring enough water. The animals were separated from the sediment by hand sieving using 0.5 mm square mesh for macrofauna as suggested by Birkett and McIntyre (1971). The sieved bottom fauna which remain above the 0.5 mm sieve was collected and preserved in 5% formalin. Further, sorting of the collected bottom fauna samples were done in the laboratory by re-sieving after washing with fresh water to remove the residual sediment and formalin. The materials remaining in the sieve were transferred to a black tray and sorted carefully by hand. All the animals in each grab sample were identified up to species level, using stereo microscope; counted and stored separately in 5% neutral formalin for further study.

Results

Totally 92 taxa of microbenthic fauna were identified of which polychaetes, crustaceans and molluscs from the major group which are represented percentage wise in Fig. 14.2 and the rest are minor groups. The macrobentic distribution, composition and numerical abundance is given below :

Coelenterates

Coelenterates were of common occurrence in the North Vembanad Lake. *Cirianthus* sp., *Obelia biscuspidata* and a species of sea anemone constituted the coelenterates. Among them, sea anemone species was represented in large numbers in station 1, 2 and 3 with a maximum number of 2240/m^2. The other two coelenterates were commonly distributed all over the area of study. The maximum number of *Cirianthus* species collected was 20/m^2 from station 12 during February 1978. *O. bicuspidata* was found distributed more numbers in stations 6 to 17. The maximum number recorded was 60/m^2 during January 1978 in Station 11.

Polychaeta

Among the macrobenthic fauna of North Vembanad Lake, the polychaetes constitute the major group throughout the period of investigation and they were present in all stations. Altogether, 33 species of polychaetes were identified, which belong to 25 different genera. Of the 33 species, 9 species were considered to be rare and 15 species were found commonly distributed in all the stations of which *Ancistrosyllis constricta, Diopatra neopalitana, Lumbriconeries latreilli, L. simplex,*

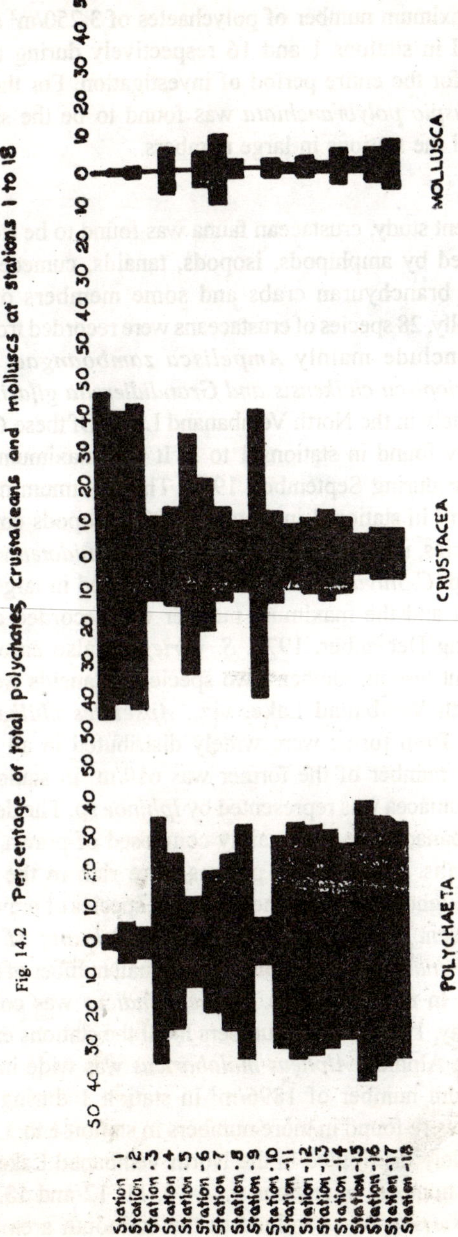

Fig. 14.2 Percentage of total polychates, crustaceans and molluscs at stations 1 to 18

Nephthys polybranchia, N. olicobranchia, Paraheteromaster tenuis, Prionaspi pinnata, P. polybranchiata were found as numerically abundant forms. The maximum number of polychaetes of 3,250/m^2 and 3,600/m^2 were recorded in stations 1 and 16 respectively during the month of March, 1978 for the entire period of investigation. For the region as a whole, *Prionuspio polybranchiata* was found to be the single species recorded in all the stations in large numbers.

Crustacea

In the present study, crustacean fauna was found to be rich in number and represented by amphipods, isopods, tanaids, cumeceans, prawns, stomatopods, branchyuran crabs and some members of the family Alphidae. Totally, 28 species of crustaceans were recorded from 22 genera. Amphipods include mainly *Ampelisca zamboangae, Corophium triaenonyx, Eriopisca chilkensis and Grandidierella gilasi.* These were distributed widely in the North Vembanand Lake. Of these *C. triaenonyx* was abundantly found in stations 1 to 3. It was maximum in station 1 with 62,420/m^2 during September, 1978. The maximum number of *G. gilasi* 20,550/m^2 in station 3 in October, 1978. Isopods compromise of only three species, namely, *Cirolana fluviatilis, Synidotea variegata* and *Anthurudea,* sp. *C. fluviatilis* was found distributed in large numbers in stations 1 to 4 and the maximum number was recorded as 489/m^2 in station 1 during December, 1978. *S. variegata* also enjoyed a wider distribution, but less in number. Two species of tanaids were collected from the North Vembanad Lake, viz., *Apseudes chilkensis* and *A. gymnophobia.* Both forms were widely distributed in all the stations. The maximum number of the former was 610/m^2 in station 10 during April, 1978. Cumacea was represented by *Iphinoe* sp. The decapod fauna of North Vembanad Lake was mainly composed of prawn, alphids and branchyurn crabs. The penaeid prawns were rich in the estuary and distributed throughout the lake. The important species of prawns collected during the present study were *Metapenaeous dobsoni; M. monoceros; M. affinis* and *Penaeus indiscus*. But, only a small number of these prawns were collected in the grab hauls. *Penaeus indicus* was collected from February to May, 1978 in large numbers in all the stations except station 16. Among the Alphids, *Alpheus malabaricus* was wide in distribution with a maximum number of 1896/m^2 in station 1 during September, 1978. Alphids were found in more numbers in station 1 to 3. Brachyuran crabs were widely distributed in the North Vembanad Lake. They were found in large numbers in stations 6, 7, 10, 11, 12 and 13. Among the crabs *Scyalla serrata* was confined to the barmouth area in station 1.

The other species of crabs like *Lithocheira, Rynchoplax* and *Viaderiana sp.* were commonly found in stations 10, 11, 12 and 13. Numerically *Viaderiana* sp., dominate the branchyuran crabs with a maximum number of 182/m^2 in station 10, during April, 1978. Stomatopoda was represented by *Squilla nepa,* a single species found confined in stations 1 to 3 and collected occasionally in the grab sample in very few numbers.

Mollusca

The molluscan fauna of North Vembanad Lake was very rich and was mainly composed of bivalves. Altogether 15 species belonging to 11 genera of the molluscan fauna were collected from 18 stations. Among the bivalves *Modiolus undulatus* and *Nuculana mauritiana* were abundant in station 3 during May, 1978. *Villorita cyprinoides* var. Cochinensis was found distributed more in number from stations 11 to 18 and the maximum number of 32/m^2 was found in station 18 during July 1978. *Babylonia spiratua* was also found commonly distributed but less in number in stations 11 to 18. *Solen* sp. collected from stations 5 to 10 and 13 to 15. Three species of *Paphia*, a bivalve was also found in stations 1 to 4 along with *Dosinia* sp. Other rare molluscs collected during the present investigation consists of *Standella* sp., 1 *Tellina* sp., *Arca* sp., *Buccinium* sp., and *Balanus* sp.

The vertebrate group was represented by only few bottom living fishes collected in the grab hauls. The gobioid fishes were represented by *Gobius microlepis, G. cristatus, Trypauchen againa* and other gobioids. *Cynoglossus semifasciatus* and *Anguilla* sp. are other fish groups represented. Apart from these group of bottom fauna, other animals collected during the present period of study includes one species of nemertine, sipunculid, echinodermata and nematoda. The nematodes and nemertines were common in occurrence, while the brittle star was found distributed few in numbers in stations 1, 2, 5, 7 and 13. The sipunculids were found in station 1, 6, 7, 8, 11, 12 and 18.

Discussion

Among the macrobenthic fauna studied, the major groups, namely, polychaetes, crustaceans and molluscs constituted about 85 to 99% of the total animals in different stations. The analysis of the distribution, composition and numerical abundance of macrobenthic fauna of North Vembanad Lake has revealed that the polychaetes dominate in 12 stations (4, 5, 8, 9, 11, 12, 13, 14, 15, 16, 17 and 18) followed by crustaceans except in station 18 where molluscs followed the polychaetes. The crustanceans were dominant in 6 stations (1, 2, 3, 6, 7 and 10) followed by polychaetes and molluscs. The molluscs were numerically less when compared to the polychaetes and crustaceans except in station 18 where the bivalve molluscs outnumbered the crustaceans. The occurrence of other groups are much less in number when compared to the major groups.

The pattern of distribution of the macrobenthic fauna of North Vembanad lake can be well understood from the marked variation found in the numerical abundance. The region near the barmouth includes stations 1, 2, 3 and 4. Here among the coelenterates, the sea anemones were distributed in large numbers in stations 1, 2 and 4 during the months of February to May, 1978. The polychaetes were mainly dominated by *Diopatra neapolitana* and *Anicstroosyllis constricta.* The crustacean fauna of their region was dominated by *Alpheus malabaricus* and *Cirolana fluviatilis.* Among molluscs, the single dominant species was that of *Modiolus undulatus* which was found distributed in large numbers during the months of March, April and May, 1978. The gobioid fishes which were recorded, show sporadic distribution in this region. Among other groups, the nematodes and nemertines were commonly distributed. Similar pattern of distribution has been reported by Pillai (1978) in this region.

The next region comprises of stations 5, 6, 7, 8 and 9 which are located away from the barmouth. In this region the coelenterates were poorly represented by *Cirianthus* sp. and *Obelia bicuspidata.* Sea anemone was not recorded because these stations are away from the barmouth. The distribution of polychaete fauna was dominated by *Paraheteromastus tenius, Prionospio polybranchiata* and *Nephthys oligobrancia* in this region. In general the crustacean fauna was found to be comparatively lesser in distribution and abundance than in the previous region. The dominating crustaceans were *Apseudes chilkensis* and *A. gymnophobia.* Among the molluscs, *Modiolus* sp. and *Sunneta scripta* were distributed fairly in large numbers. The notable feature in this region was the occurrence of large numbers of *Nuculana mauritiana* during the month of April, 1978 in station 8. The third area includes stations from 10 to 15 which are located in the mid-region of the North Vembanad Lake. This region, more or less shows similar distribution and abundance of coelenterates, polychaetes and crustaceans with little variations. Among crustaceans, the branchyuran crab *Viaderiana* sp. was observed to be dominant in station 10 to 13. The molluscan fauna was represented by *Nuculana* sp. in stations 10 and 11 but comparatively less in number than in the previous region *Villorita cyprinoides* var *cochinensis* was found recorded in large numbers in stations 12 to 15.

The last region comprises of stations 16, 17 and 18 where the coelenterates were very poorly recorded than in all the other regions, because this region is far away from the barmouth. The polychaete fauna of this region was abundantly represented by *Prionospio polybranchiata*. The dominating crustaceans were *Briopisca chilkensis and Gammares* sp. The molluscs include large number of *Villorita cyprinoides var. cochinesis* in this region in stations 17 and 18 during December, 1978.

In general, the distribution of sea anemones was restricted to barmouth area. The polychaetes were found commonly, in all the regions. The maximum species representation of 26 species were noticed in station 1 and 2 and the minimum number of 13 species were recorded in station 15. The crustaceans were found in large number in the barmouth area in stations 1 to 3 and in the mid-reaches of the North Vembanad Lake in stations 10 to 12. The molluscs were found distributed in all the regions but they are patchy in existence.

Acknowledgement

The author is thankful to Dr. R. Damodaran, Professor of Marine Biology for his guidance and for the UGC for financial assistance. Thanks are also due to the Director, School of Marine Sciences and to the University of Cochin for providing the necessary facilities.

References

Anzari, Z.A., (1977) Macrobenthos of Cochin backwaters, Bull. Natn. Inst. Oceanogr. V. 10, pp. 169-71.

Batcha, S.M.A. and Damodaran, R., (1971) Impact of Thanneermukkam bund and Idukki Hydro-electric Project on the changes of salinity characteristic of the Vembanad Lake (South India). Arch. Hydrobiol. Beith Ergebn. Limnol. V. 28, pp. 193-200.

Birkett, L. and Mc Intyre, (1971) Treatment and sorting of samples, in Methods for the study of Marine Benthos. IPB Hand Book, No. 16, Blackwell Scientific Publications, Oxford and Edinburg, pp. 156-68.

Desai, B.N. and Krishnmkutty, M., (1967) Studies on the benthic fauna of Cochin backwaters, Proc. Ind. Acad. Sci. V. 55, pp. 123-42.

Kitamori, R., (1963) Studies on the benthic communities of littoral sea in the Seto-Island sea and adjacent waters. Bull. Naikal Reg. Fish. Res. Lab., No. 21, pp. 1-90.

Kurian, C.V., (1972) Ecology of Benothos in a tropical estuary. Proc Indian Nat. Sci Acad. Part B, V. 38(3-4), pp. 156-63.

Kurian, C.V., Domodaran, R. and Antony, A., (1975) Bottom Fauna of the Vembanad Lake. Bull. Dept. Mar. Sci. Univ. Cochin, V. 7, pp. 987-94.

Miyadi, D., (1941) Ecological survey of benthos of the Ago-Wam. Ibid., V. 20, pp. 169-80.

Pillai, N.G.K., (1977) Distribution and seasonal abundance of the macrobenthos of the Cochin backwards. Indian J. Mar. Sc., V. 61, pp. 1-5.

Pillai, N.G.K., (1978) Macrobenthos of a Tropical estuary, Ph. D. Thesis, University of Cochin, 133 pp.

15

Evaluation of Water-Quality and Trophic Status of Sagar Lake

GUNWANT JOSHI

AND

A.D. ADONI

Introduction

Trophic state of any aquatic ecosystem basically refers to its soluble nutrients and resultant biotic productivity along its natural succession from oligotrophic (less productive) to eutrophic (highly productive) state. Surprisingly, however, there is no generally accepted unique definition of a lake's trophic status, nor is there a quantitative measure (Imboden and Gachter, 1978).

In practice, the term trophic status is very often, and also vaguely, referred to in relation to a host of water quality parameters. But evolving a generally applicable trophic status index, in quantitatives terms, based on these parameters, is difficult because of the lack of sufficient information regarding complexity of interactions amongst various environmental factors and their effect on fresh water organic production (Brook, 1965). On the other hand, the efforts of classifying the water bodies along a trophic gradient on the basis of biological indices alone, may also be misleading. The concept of phytoplanktonic quotients has been most remarkably worked out by Nyggard (1949) for Danish lakes. But out of the five different quotients suggested by him, only the compound index has been found to be useful and applicable to waters over a wide geographical range (Hutchinson, 1967). Although, the quotient was inapplicable in North America (Rawson, 1956), Gunale and Balkrishnan (1981) have reported to it be fairly sensitive, reliable and thus suitable for tropical water bodies of India.

In present study, an attempt has been made to assess trophic status of the Sagar lake on the basis of different physocio-chemical and biological parameters.

Material and Methods

Water samples were collected from two stations (one littoral and one limnetic) in the lake, twice a month from different depths at 9 a.m. and

10 a.m. respectively. Most of the routine physocio-chemical analysis were done in the field itself within 20 minutes of collection. The nutrients were analysed in the laboratory from the preserved samples within 24 hours of collection. BOD samples and plankton samples for chlorophyll estimations were transported to laboratory immediately in sterile dark containers where they were further processed within an hour of collection. Standard Methods (APHA, 1976) were used for chemical analysis. Analysis of dissolved carbodhyrates was done using anthrone method following Golterman *et al.* (1978). BOD and COD were done by the methods suggested by Adoni *et al.* (1985). The plankton community structure was studied from the composite column sample obtained by mixing plankton samples of all the depths from surface to bottom in deep waters. The planktonic collections were condensed by sedimentation technique using Lugol's iodine as preservative. The drop count method (APHA, 1976) was used for plankton enumeration. Chlorophyll 'a' content was estimated colorimetrically from the acetone extracts of fresh planktonic samples separately for the total and the net plankton filtered over No. 25 bolting silk. The nannoplanktonic chlorophyll 'a' content was estimated by subtracting chlorophyll 'a' values for net plankton from those of total phytoplankton. Primary productivity of the phytoplanktonic community was estimated by the light and dark bottle method of Gaarder & Gran (1927) by short and long exposure technique (Joshi and Adoni, 1987). Compound quotient (Nyggard, 1949) was used for analysis of algal associations. Zooplankton samples were collected over No. 25 bolting silk and were fixed by adding one ml of 5% formaline + few drops of glycerine to 20 ml. of plankton concentrate.

Results and Discussion

Table 15.1 shows the summarised values of physico-chemical and biological quality of water as annual average and range of variation. Tables 15.2 and 15.3 show the phytoplankton and zooplankton species encountered at Sagar lake whereas Table 15.4 shows the monthly changes in the community characteristics of Sagar lake plankton during present study. Table 15.5 shows some trophic parameters of Sagar lake compared with other two Indian lakes and with the standard ranges of trophic status suggested by some workers.

A careful study of Table 15.1 reveals that the range of fluctuations of almost all the parameters is significantly wider. The water quality parameters such as low Secchi transparency, high pH and high calcium, nitrate and phosphate contents indicate the eutrophic nature of the water

Table 15.1 : Physico-chemical and Biological Characteristics of Sagar Lake

Parameter (Unit)		*Annual average*	*Range*
Waterlevel	(m)	3.88	2.0-5.5
Water temperature	(°)	22.60	18.0-29.5
Secchi depth	(cm)	56.00	16-104
pH	(Units)	8.7	7.5-9.5
Free CO_2	(ppm)	Rare	0-2
Bicarbonates	(ppm)	63	24-124
Carbonates	(ppm)	37	6-72
Dissolved O_2	(ppm)	6.4	1.8-9.2
Chlorides	(ppm)	48	32-95
Ca^+ hardness	(ppm)	52	35-63
Total hardness	(ppm)	95	84-106
Nitrates	(ppb)	161	20-383
Phosphates	(ppb)	60	14-135
Carbohydrates	(ppm)	5.4	2-19.7
BOD_3 27°C	(ppm)	10.1	8.0-11.4
COD	(ppm)	45.3	20-60
Chl. 'a' for Nannoplankton	(mg/m^3)	128	25.5-370
Gross Pri. Productivity			
24 Hrs. exposure	($mg\ C/m^3/d^1$)	2,100	980-3,680
3 Hrs. exposure	($mg\ C/m^3/d^1$)	5,200	1,350-7,600

body as is further evidenced by the higher biological productivity in terms of dissolved organic carbon, chlorophyll 'a' and the gross primary production estimates. Presence of high chloride concentration in water indicates the sewage pollution of animal origin. The changes in water quality over more than a decade indicate clearly that Sagar lake is showing an escalating degree of eutrophication and pollution (Joshi and Adoni, 1990).

Table 15.2 : Phytoplankton species occurring at Sagar lake

CHLOROPHYCEAE	CYANOPHYCEAE
Elakatothrix gelatinosa	*Microcystis aeruginosa*
Golenkinia radiata	*Merismopedia* sp.
Micractinium sp.	*Gleocapsa* sp.
Schroederia indica	*Chroococcus* sp.
Pediastrum simplex	*Oscillatoria* sp.
P. ovatum	*Spirulina* sp.
P. duplex	*Anabaena* sp.
P. muticum A. spiroides	*A. Spiroides*
Coelastrum microporum	*Anabaenopsis* sp.
Pachycladon sp.	
Tetraedron minimum	**EUGLENOPHYCEAE**
T. gracile	*Trachelomonas* sps.
Treubaria sp.	*Phacus* sp.
Polyedriopsis spinulosa	**BACILLARIOPHYLEAE**
Ankistrodesmus sp.	*Melosira granulata*
Tetrastrum sp.	*Cyclotella* sp.
Scenedesmus armatus	*Fragilaria* sp.
S. accuminatus	*Synedra* sp.
S. quadricauda	*Gomphonema* sp.
S. abundans	*Cymbella* sp.
S. bijugatus	*Epithemia* sp.
S. bernardii	*Rhopalodia* sp.
Closterium actum	*Nitzschia* sp.
Cosmarium circulare	
C. depressum	

The phytoplankton included the members of Cyanophyceae (8 genera, 9 sps.), Chlorophyceae (17 genera, 26 sps.), Bacillariophyceae (9 genera) and Euglenophyceae (2 genera) of which only the first three groups were of quantitative importance. The values of compound quotient were found to be between 4 to 13.7, highlighting the eutrophic state of the lake. Nyggard (1955) himself stated that the compound quotient values for low productive waters is generally less than 1.2; values up to 3.0 indicated high eutorophic nature and values 3.0 to 13.0 indicated definitely

Table 15.3 : Zooplankton species occurring in Sagar lake

PROTOZOA	ROTIFERA
*Arcella discoides**	*Brachionus angularis*
*Centropyxis aculeata**	*B. calyciflorus*
*Difflugia urceolata**	*B. caudatus*
*Epistylis plicatilis**	*B. forficula*
	B. falcatus
CLADOCERA	*B. guadridentata*
*Sida crystallina**	*Anuraeopris fissa*
Daphina hylina	*Keratella tropica*
*D. carinata**	*K. cochlearis*
*D. longispina**	*K. ticinesis*
Moina brachiata	*Notholca* sp.*
Moinodaphania sp.	*Platyias quadricornis**
Chydorus sp.*	*Lepadella ovalis**
Bosmina longirostris	*Trichocerca longiseta*
*Ceridaphnia reticulata**	*T. multicrinis**
	*Lecane luna**
COPEPODA	*Monostyla lunaris**
Cyclops sp.	*Gastropus stylifer*
Nauplii	*Polyarthra* sp.
	Filinia opoliensis
	F. longiseta
	Hexarthra sp.
	Conochilus unicornis
	Rotaria sp.*

* Organisms of rare occurrence recorded only occasonally

eutrophic characteristics. Gunale and Balakrishnan (1981) found the compound quotient values for rivers varying from 2.15 to 13.75 and Valecha *et al.* (1987) found to be between 10.0 to 19.0 for Lower lake of Bhopal. Further when we compare the values for organic carbon, planktonic biomass as chlorophyll 'a', primary production values and the types of dominant algal associations with that of two lakes of Hyderabad (Mohan, 1980) and the classification criterai suggested by Likens (1975), the eutrophic nature of the system is confirmed.

Another interesting inference comes from the observations on zooplankton with a total of 27 genera and 38 species represented by rotifers, protozoans, copepods, cladocerans and ostrocods. Rotifers represented by 15 genera and 24 sps accounted for about 47 to 97% of

Table 15.4 : Changes in commonly characteristics of Plankton at Sagar lake

Parameters	*July*	*Aug.*	*Sept.*	*Oct.*	*Nov.*	*Dec.*	*Jan.*	*Feb.*	*Mar.*	*Apr.*	*May*	*June*
Compound quotient	13.3	8.3	8.0	12.0	*	*	4.0	4.0	8.0	6.3	9.0	13.7
% dominance of algal sps	N 78 Sc 14 G 4.6 Me 3.5		M 78 N. 18	M 59 N 14	Mi 52 M 19	M 26	M 57 Mi 23	M 48 M 18 S 15 A 6.2	Mi 37 S 26 M 7.2 Sc 70	Mi 63 Sc 23 E 6.6 S. 9.8	N 30 M 20 S 18	N 17 M 16 Mi 9.9
Rotifer density												
Year 1970-71	03	09	31	21	47	153	115	117	48	281	305	291
Year 1978-79	304	428	364	264	300	312	1032	495	572	400	416	304
Present study	3,037	440	474	481	460	656	827	4590	9897	4180	696	556

* Compound quotient could not be calculated as no desmids were observed

Index to algal species : A =Anabena, E=Epithemia, G = Gloeo capsa, M = Melosira,

Me = Merismopedia, Mi = Microcystis, N = Nitschia, S = Synedra,

Sc = Scenedesmus

Rotifer density studies : 1970-71 (Adoni, 1975) ; 1978-79 (Awatramani, 1980)

Table 15.5 : Some Trophy Parameters Compared with Other Lakes

Lake/Type	*Parameters*				*Reference*
	Organic C $mg\ l^{-1}$	*Chlorophyll 'a'* $mg\ m^{-3}$	*Productivity* $mg\ Cm^{-3}d^{-1}$	*Dominant phytoplankton*	
Osman Sagar	1.8	150	911	Cyanophyceae Bacillariophyceae	Mohan (1980)
Mir Alam Lake	5.7	1,780	3,836	Cyanophyceae	Mohan (1980)
Sagar Lake	5.4	137.5	2,100* 5,200*	Cyanophyceae Bacillariophyceae	Present study
Oligotrophic	< 1-3	0.3-3	20-100	Chrysophyceae Cryptophyceae Dinophyceae Bacillariophyceae	Likens (1975)
Mesotrophic	< 1-5	2-15	100-300		
Eutophic	5-30	10-500	> 300	Bacillariphyceae Cyanophyceae Chlorophyceae Euglenophyceae	

* 24 Hrs (long exposure) technique

** 3 Hrs (short exposure) technique

the total zoolplanktonic number during the study period. Since long, the rotifers have been used a bioindicators of water quality because of their higher species diversity and cosmopolitan distribution. *Brachionus angularis, B. calyciflorus, B. quadridentata* and *B. plicatilis* have been reported to be eutrophic forms (Arora, 1966 and Datta & Bandyopadhyaya, 1984). *Keratella cochlearis, Anuraeopsis, Polyarthra* and *Filinia longiseta* are also reported to be eutrophic forms (1966) but there were instances when these forms are recorded in oligotrophic waters, thus not only mere presence or absence of species but also its population density should be assessed (Chaurasia, 1987). *K. cochlearis* known to be a eutrophic species is perhaps the most abundant zooplankton in Sagar lake at times attaining the population density up to 65% of the total zooplanktonic numbers. Further, interesting observation on zooplankton is that over the years population density of rotifers is steadily increasing, indicating that the rotifer population increases with escalating eutrophication.

Acknowledgements

The present work was carried out under Grant No. 26/6/78 from Department of Environment to ADA in form of MAB Project. Grateful thanks are due to Dr. S.K. Chaurasia and Dr. A.K. Vaishya for help rendered in plankton enumeration.

References

Adoni, A.D., Joshi, Ghosh, Chaurasia, S.K., Vaishya, A.K., Yadav, M. and Verma, H.G. (1985), Work Book on Limnology, Pratibha Publishers, Sagar. 216 pp.

Adoni, A.D. (1975) Microbology, of Sagar Lake, Ph. D. Thesis, University of Sagar.

A.P.H.A., (1976) Standard Methods for Examination of Water and Waste Water. 14th Edition. APHA. AWWA. WPCF, Washington, 1193 pp.

Arora, H.C., (1966) Rotifera as indicator of trophic nature of environment. Acta Hydrobiologia XXVII Fasc. 1-2.

Awatramani, G.M., (1989) Limnological studies of Sagar lake, Sagar. Ph. D. Thesis, University of Saugar, 294 pp.

Brook, A.J., (1965) Planktonic algae as indicators of lake types with special reference to the Desmidiaceae. Limnol. Oceanogr. V. 10, pp. 403-11.

Chaurasia, S.K. and Adoni, A.D., (1987) Rotifers as indicators of eutrophication. In Perspectives in Hydrobiology (ed). Rao, K.S. and S. Shrivastava, Sec. IV (28), pp. 147-49.

Datta, N.C., and Bandyopadhyay, B.K., (1984) Rotifers as bio-indicators of the state of aquatic environments. procs. Int. Symp. on Biological Monitoring of the State of Enviromental. 11-13 Oct. 1984, New Delhi (Abstract).

Golterman, H.L., Clymo, R.S. and Ohnstad, M.A.M. (1978) Methods for physical and chemical analysis of fresh waters. Blackwell Scientific Publications, Oxford, 213 pp.

Gunale, V.R., and Balakrishnan, M.S. (1981) Biomonitoring of eutrophication in Pavana, Mula and Mutha rivers flowing through Poona. Ind. J. Environ, Hlth. V., 23(4), pp. 316-21.

Hutchinson, G.E., (1957) A Treatise on Limnology. Vol. I, Geology, Physics and Chemistry; Wiley, New York, 1015 pp.

Hutchinson, G.E., (1967) A Treatise on Limnology. Vol. II. Introduction to lake biology and Limnoplankton. Wiley, New York, 1115 pp.

Imboden, D.M. and Gachter R., (1978) A dynamic lake model for trophic state predication. Ecol. Modelling. V. 4. pp. 77-98.

Joshi, G., (1987) Diurnal Studies of physico-chemical properties of limnological importance. Ph.D. Thesis, Dr. Harisingh Gour Vishwavidyalaya, Sagar, 204 pp.

Joshi, and Adoni, A.D., (1987) 'Primary productivity of a natural lake by short and long exposure methods using light and dark bottle technique. Geobios New Reports. V 6(1), pp. 87-88.

Joshi, and Adoni, A.D., (1990) Progression of water quality at Sagar lake: A case for escalating eutrophication and Pollution. Bpl. Univ. Res. J., V. 2, pp. 19-21.

Likens, G.E., (1975) Primary production of inland aquatic ecosystems. In the primary productivity of the Biosphere (ed.). Lieth, J. and R.H. Whittakar, Springer-Verlag, New York.

Mohan, J. Satya, (1987) Trophic status of two lakes of Hyderabad, India. In Perspectives in Hydrobiology. (ed.) Rao, K.S. and S. Shrivastava, Sec. IV (38), pp. 193-99.

Nyggard, (1949) Hydrobiological studies of some Danish ponds and lakes II. The quotient hypothesis and some new or little known phytoplankton organisms. Kal. Danske Videnskab. Selskab. iol. Skrifter V. 7(1), 293 pp.

Nyggard, (1955) As cited by Hutchinson, C.E. 1957.

Pearsall, W.H., (1932) Composition of the phytoplankton in relation to dissolved substances. J. Ecol. V. 20, pp. 241-61.

Rawson, D.S., (1956) Algal indicators of trophic lake types. Limnol. Oceanogr. V. 1, pp. 18-25.

Rao, S.V., Rao, M.S., Singh, V.P. and Mall, L.P., (1981) Algal diversity in relation to water quality in an Indian river. Int. J. Environmental studies, V. 16, pp. 101-05.

Somashekar, R.K., (1985) Studies on water pollution of river Cauvery. Proce. Nat. Symp. Pure & Appl. Limnology. (ed.) A.D. Adoni, Bull. Bot. Soc. Sagar, V. 32, pp. 145-49.

Thunmark, S., (1945) As cited by Hutchinson, G.E. 1967.

Valecha, V., Trivedi, R. and Bhatnagar, G.P., (1987) Biological assessment of trophic status of Lower lake of Bhopal. Poll. Res. V. 6 (3 & 4), pp. 91-93.

16

Fungal Digestion of Deprotenized Leaf Juice from two Aquatic Weeds for SCP Production and BOD Reduction

S. Chanda

Introduction

The damaging aspect of aquatic weed growth on the environment have far been emphasized. Eradication of these weeds or to maintain the plant population at managable level by mechanical, chemical or biological mean requires recurring efforts involving high machine and labour inputs or chemicals, pathogens or insects indigenous to aquatic weeds which possess the virtues of high specificity and lethality without adverse side effects. At present there is a striking paradox involved when efforts are directed towards the exploitation of the fresh water ecosystem as a potential crop for leaf protein production where acute protein calorie malnutrition prevails. Studies on the utilization of aquatic weeds as animal feed, human food, soil additives and fuel production are now receiving increased attention as this would greatly offset their high cost of removal (NAS, 1984). Exploitation of this fresh water ecosystem as potential crop for leaf protein has gained considerable momentum in recent years through the valuable work of Boyd, 1968; Matai, 1976; and Bandyopadhyay, 1989.

During large scale manufacture of leaf protein from aquatic plant sources, considerable amount of liquid deproteinized juice (DLJ) or brown juice is produced as by-product. The by-product as effluent of this food processing industry suffers from the problem of disposal. Due to the presence of appreciable amount of recycleable nutrients and also having high chemical and biochemical oxygen demand, it will be rather unwise to dispose it into any water body on the stand point of pollution, treating waste by microbiological means may serve as an attractive and inexpensive method prior to its disposal into any environmental system.

The investigation was undertaken to determine the ability of two filamentous fungi to grow in deproteinized brown leaf juice from two aquatic weeds for single cell protein production and BOD reduction as it is a well known fact that the filamentous fungi play a significant role in reduction of pollutant parameters of a waste and simultaneously mycelium offers a good source of protein (Church *et al.*. 1973).

Material and Methods

The two plants *Allmania nodiflora* and *Marsilea quadrifolia* were collected from different localities in and around Calcutta. The plant material was collected and placed in polythene bags and transported to the laboratory. Juice was extracted by the methods using pulper (Davys and Pirie, 1969) and press (Davys, Pirie and Street, 1969). Protein was coagulated at 80°C, the fluid left after heat coagulation of protein is brown leaf juice or deproteinized leaf juice (DLJ). Brown juice was collected by centrifugation at 20,000 rpm and stored by autoclaving at 10 lbs pressure for 10 minutes for chemical analysis and fungal propagation.

The two fungi *Fusarium pallidoroseum* and *Gliocladium roseum*, the two natural contaminants of juices were further exploited in the fermentation experiments. The fungi were maintained in Potato Dextrose-Agar slants. In a typical experiment, 250 ml sterile conical flasks, each containing 50 ml of specific DLJ sample, were used for this study. pH of the medium was 5.5. Incubation temperature was 28°C. Fermentation was conducted in still culture for 7-9 days to get maximum mycelial yield. The mycelial felt was harvested after incubation and filtered through a pre-weighed whatman filter paper No.1, washed several times with distilled water and dried in a hot air oven at 70°C for over night. The nitrogen content of the wastes and fungal biomass was determined by micro-kjeldahl method and crude protein content (NX 6.25) of biomass was then calculated. The reducing sugar content was calculated as anthrone positive glucose equivalent. Lipid was extracted by chloroform, methonal mixture 2:1. Ash content was determined by keeping the samples at 550°C in Muffle furnace for six hours COD (Taras *et al.*, 1971) was measured before and after fermentation by diluting the sample (1:1000) with redistilled water, organic and oxidizable substances were estimated, for BOD (Ballinger, 1979) determination the samples were diluted to 1:1, 1:10, 1:100, 1:1000 ratio as required. The principle underlying in routine BOD test consists in determining the dissolved oxygen (DO) consumed during incubation for 5 days the BOD of the diluted samples was calculated from dissolved oxygen consumed.

Results and Discussion

A very serioius problem arising from modern technology is pollution. Land, water and air pollution due to municipal, agricultural and industrial wastes are generally controlled or reduced with subsequent reduction of COD and BOD by harvesting a number of microbes such as photosynthetic bacteria (Kobayashi, 1975), *Spirulina* (Kosaric *et al*., 1974), filamentous fungi (Church, *et al*., 1973) and yeast. In this study *Fusarium pallidoroseum* and *Gliocladium roseum* were employed to reduce the COD and BOD values of the wastes through fermentation in still culture for 7-9 days to get maximum mycelial yields. In due course of study initial and final concentrations of carbohydrate and nitrogen sources were calculated to investigate the percentage of reduction of two major nutrients required for harvesting the protein rich biomas. The BOD and COD values were also seen to be decreased progessively due to fermentation for a period of 7-9 days.

The role of fungi in pollution control and microbial protein production is well documented. Treatment of many food processing wastes by fungal means serves as an attractive and inexpensive means prior to its disposal into any environmental system. Hang *et al*. (1975) used *Aspergillus niger* to reduce the BOD from 22,500 to 900 mg/L by harvesting biomass. Hang and Woodams (1977) employed *Aspergillus faetidus* to reduce BOD from 30,000 to 7,000 mg/L. The recovered myceilum contains 40% protein. Kapoor *et al*., (1981) harvested a filamentous fungi that was *Rhizopus nigricans* in a by-product from molasses distillery unit and showed that the Chemical Oxygen Demand of the by-product was reduced by 29.8% due to fermentation. It has been noticed that deproteinized leaf juices from both *Allmania* and *Marsilea* are loaded was essential nutrients for microbial propagation and as a matter of fact both Chemical and Biochemical Oxygen Demands of the wastes are quite high (Table 16.1). When both the deproteinized juices were used as fermentation media *Fusarium pallidoroseum* consumed 98% carbohydrate and 75-78% nitrogen and reduced 79-81% Chemical Oxygen Demand and 95-98% Biochemical Oxygen Demand and subsequently yielded biomass that was 7.2-8.5g/L of fermented wastes. Dried Biomass contains 30-32% of protein (Table 16.2a) when *Gliocladium roseum* was employed in the DLJ substrates, it consumed 97-98% available carbohydrate, 70-73% nitrogen and progressively reduced 70-75% Chemical Oxygen Demand and 90-93% Biochemical Oxygen Demand. It yielded 9.6-10.5g dried biomass/L of fermented waste. Dried biomass contain 26-28% protein (Table 16.2b).

Table 16.1 : Characteristics of deproteinized leaf juices (DLJ) from aquatic weeds

Characteristics mg/l	*Allmania nodiflora*	*Marsilea quadrifolia*
Dry weight	45,000	40,000
Ash	21,750	23,350
Reducing sugar	10,250	8,800
Nitrogen	8,100	6,200
Lipid	2,500	1,525
COD	25,100	21,750
BOD	20,225	17,125

Table 16.2 : Nutrient utilization COD and BOD reduction yield and protein content of fungal biomass in deproteinized leaf juice media by (a) *Fusarium* (b) *Gliocladium*

Source of DLJ	*Reducing sugar utilization %*	*Nitrogen utilization %*	*COD reduction %*	*BOD reduction %*	*Yield of biomass g/l*	*Protein content of biomass %*
(a) *Allmania nodiflora*	98	75	79	98	8.5	32
Marilea quadrifolia	98	78	81	95	7.2	30
(b) *Allmania nodiflora*	97	70	72	90	10.5	28
Marsilea quadrifolia	98	73	75	93	9.6	26

Worgan and Wilkins (1977) harvested three filamentous fungi *Fusarium semitectum, Trichoderma viride* and *Aspergillus oryzae* in deproteinized juice from lucerne and those fungi removed 65-78% of COD of DLJ. It has been observed recently (Chanda *et al.*, 1990) that either COD or BOD of deproteinized juices were progressively reduced by 40-70% and 78.65-93% respectively due to the growth of a good number of filamentous fungi. The carbonaceous and nitrogenous part of the by-products were progressively consumed for the production of biomass (0.44-3.4g/100 ml of waste) and the protein of dry harvested biomass was 19.37-29.70% depending upon the waste or fungi used. Church *et al.* (1973) employed a number of filamentous fungi such as *Trichoderma, Gliocladium* and *Geotrichum* to reduce to COD and BOD of corn and pea processing wastes by (81-88%) and (95-96%) respectively. All these above results harmonises the recent findings. Church and Nash (1970) pointed out the use of fungi imperfecti in waste treatment as the protein quality of the fungal mass to be equivalent to that of soya meal. Both *Fusarium* and *Gliocladium* belong to fungi imperfecti may serve as an alternative protein source to meet the global gap of protein calorie malnutrition.

References

Anonymous., (1984) Making aquatic weeds useful : Some perspective for developing countries. Report on an Adhoc Panel of Advisory Committee on Technology Innovation Board on Science and Technology for International Relations, National Academy of Sciences, Washington D.C. pp. 175.

Ballinger, D.G., (1979) Methods for Chemical Analysis of Water and Wastes. Environmental Monitoring and Support Lab. US Environmental Protection Agency, Ohio, U.S.A.

Bandyopadhyay, A., (1989) Utilization of aquatic weeds with stress on their nutritive value. Ph.D. Thesis, Dept. of Agriculture, University of Calcutta.

Boyd, C.E., (1968) Fresh water plants: A potential source of protein. Econ. Bot. (USA) V. 22, pp. 259-368.

Chanda, S., Chakrabarti, S. and Matai, S., (1990) Reduction of pollutant parameters of leaf protein concentrate by-product by microbial fermentation J. Ferment. Bioeng. V. 69(5), pp. 308-10.

Church, B.D., Erickson, E.E. and Widmer, C.M., (1973) Fungal digestion of food processing wastes. Food. Tech. V. 27, pp. 36-42.

Church, B.D. and Nash, H.A., (1970) Fungi imperfecti in waste control. Water Pollution Control Research. Series 120 60 EHT. US. Govt. Print Off. Washington D.C.

Davys, M.N.G. and Pirie, N.W., (1969) Laboratory scale pulper for leafy plant material. Biotechnol. Bioeng. V. 11, pp. 517-28.

Davys, M.N.G., Pirie, N.W. and Street, G., (1969) A laboratory scale press for extracting juice from leaf pulp. Biotechnol. Bioeng. V. 11, pp 529-38.

Hang, Y.D., Splittstoesser, D.F. and Woodams, E.E., (1975) Utilization of brewery spent grain liquor by *Aspergillus niger*. Appl. Microbiol V. 30(5), pp. 879-81

Hang, Y.D., Woodams, E.E. (1977) Baked bean waste : A potential substrate for producing amylase. Appl. Environ. Microbiol. V. 33(6), pp. 1293-94.

Kapoor, B.K., Bhatia, O.P. and Misra, A.K., (1981) Mould protein from molasses distillery spent waste, Ind. J. Microbiol. V. 21, pp. 283-87.

Kobayashi, M., (1975) Role of photosynthetic bacteria in foul water treatment. Prog. Wat. Technol. V. 7, pp. 309-15.

Kosaric, N., Guyen, H.T. Bergougnou, M.A., (1974) Growth of *Spiruina maxima* algae in effluent from secondary waste water treatment plants. Biotechnol. Bioeng. V. 16, pp. 881-96.

Matai, S., (1976) Protein from aquatic weeds. In Aquatic weeds in South East Asia (Eds. C.K. Varshney and J. Rozoska). W. Junk, the Hague, Netherlands, pp. 369-73.

Taras, M.J., Greenberg, A.E., Hoak, R.D. and Rand, M.C., (1971) Standard methods for the examination of Water and Waste Water, Ed. Am. Pub. Health, Assoc. Washington, D.C.

Worgan, J.T. and Willkins, R.J., (1977) Green crop fractionaction Occasional Symposia, 9 Br. Grassel Soc. Maidenhead.

17

On the Phytoplankton Flora of a Few Temple Tanks in Madras and their Unique Phycobiocoenoses

N. JEEJI BAI
AND
D. LAKSHMI

Introduction

Heleoplankton, characteristic of small fresh water pools, puddles, etc., as different from the limnoplankton of large lakes and reservoirs, has received increasing attention because of its rich floristic diversity (Thomasson, 1979). In Tamil Nadu small ponds, rainwater pools and puddles have been studied since long for the quick succession of their interesting algal flora (Iyengar, 1920; 1939; Ganapati, 1941; Ganapati *et al.*, 1953; Chako, 1972). The reviews and monographs on different algal groups which have been published in India so far are mainly compendia of such investigations.

Temple tanks represent unique ecosystems as there is practically no run-off from sides unlike natural ponds. As they are enclosed by low walls, the changes in water chemistry are mainly due to bottom mud and silt besides rainwater dilution, the underlying spring and human activity (washing, bathing, etc.). Earlier investigations on temple tanks were concerned with the water blooms and hydrological conditions causing them (Iyengar 1939; Ganpati *et al.*, 1953). The present study was undertaken to understand their flora, the quick changes occurring in them after rains and the differences in their physico-chemical and floristic composition.

Materials and Methods

Samplings were made once in October and November 1978 and twice a month from January to March 1979 in Gangadhareswarar and Vadapalani temple tanks and twice a month from February to April in Parthasarathy temple tank. Water samples were collected with due precautions after taking the temperature readings. Phytoplankton was then collected by filtering 40-60 litres of water through a plankton net of about 30 µm pore size. Water analysis data of only those samples which showed differences have been given in Table 17.1 along with relevant references. Phytoplankton was examined in the living state and after fixing in 4% formalin for taxonomic determination using the relevant manuals.

Hydrobiology of the three tanks investigated

1. Gangadhareswarar Temple Tank

This tank is situated in the heart of the city of Madras. It is large (68 metres × 64 metres) with about 4 metres deep water at the time of study. Access is only on two sides with steps on one of them, the other two sides being enclosed by walls. Part of the tank is shaded by trees.

Aquatic macrophytes such as *Ceratophyllum* (abundant), *Nymphaea, Eichhornia* and grasses occurred with blooms of *Wolffia* and *Lemna* in February and March 1979. Water was colourless, somewhat alkaline having some free CO_2 and no carbonate and a fairly high level of bicarbonate. The N/P ratio dropped from 3.6 in November to 0.7 in February and March. Other parameters did not vary significantly (Table 17.1).

A diverse phytoplankton flora occurred in this tank with 32 taxa comprising 13 species of Chlorococoales and 10 of blue greens, 4 of Euglenophytes, 3 each of desmids and diatoms and one Xanthophyte (Table 17.2). A majority of these were observed in small numbers throughout the period of study. Noteworthy features of the phycocoenosis were the simultaneous occurrence of cyanophytes, diatoms, desmids and Euglenophytes in fair numbers; the abundant appearance of *Rhabdoderma minima* in February and its persistence till March and the presence, also in good numbers, of a Thecamoeba, *Arcella sp.* throughout the study period.

During October-November 1978 when there was intermittent rains the number of species and their abundance was low. This was probably because of the low profile of phytoplankton during the long summer preceding the rains and the sudden turbulence and dilution after that. From January the diversity and individual abundance of the taxa steadily increased up to March 1979, the stable conditions facilitating the quick reestablishment of the taxa (Table 17.2).

2. Vadapalani Temple Tank

This tank is situated a little away from the centre of Madras, measures 68 metres × 68 metres with 4.5 meters deep water at the time or study and has access with steps on all four sides. There is no tree cover at all around the tank.

Table 17.1 : Water Analysis Data of the Three Temple Tanks Investigated

Parameters	*Gangadhareswarar Temple Tank*					*Vadapalani Temple Tank*					*Parathasarthy Temple Tank*			
	Dates of Sampling					*Dates of Sampling*					*Dates of Sampling*			
	21.11.78	2.1.79	17.1.79	2.2.79	1.3.79	23.11.78	3.1.79	18.1.79	1.2.79	1.3.79	16.2.79	1.3.79	17.3.79	2.4.79
1. Temperature °C	23.4	23.0	23.1	23.5	25.0	26.0	25.0	25.5	27.5	28.8	27.5	29.0	30.0	30.5
2. pH (electrode)	7.9	8.0	8.1	8.2	8.6	7.7	8.9	9.9	9.0	8.8	7.85	8.6	7.4	8.25
3. Conductivity-umhos/Cm³ (electrolytic)	-	833	105.4	1110	1221	-	444	472	527	527	361	389	416	444
4. Dissolved Oxygen-ppm (modified Winkler, Theroux *et al.*, 1943)	2.0	2.0	9.5	4.8	6.4	4.0	8.0	12.5	24.0	4.8	19.4	9.6	3.6	4.0
5. Free Carbon dioxide ppm (as above)	19.7	15.0	13.0	22.0	17.4	11.0	Nil	Nil	Nil	Nil	24.0	10.0	12.0	9.5
6. Carbonate-ppm (as above)	Nil	Nil	Nil	Nil	Nil	Nil	19.4	100	140	30	Nil	Nil	Nil	Nil
7. Bicarbonate-ppm (as above)	913	608	450	465	480	335	213	65	20	135	110	120	140	166
8. Phosphate-µg PO_4-P/l (Barnes, 1959)	12.5	28.0	36.5	32.0	38.0	25.5	24.0	24.0	25.0	25.0	12.5	25.0	22.0	29.0
9. Nitrate-µg NO_3-N/l	46.0	60.0	53.0	24.0	25.5	31.5	42.5	51.5	34.0	26.0	28.0	21.0	20.0	20.0
10. Dissolved Organic Matter-ppm (ALPHA, 1962)	-	9.35	9.73	9.80	10.0	-	9.31	10.0	10.0	10.0	8.43	10.0	10	10.0

Table 17.2 : Changes in Phytoplankton Community in Gangadhareswarar Temple Tank

		Date of Sampling				
	Phytoplankton	*21.11.78*	*2.1.79*	*17.1.79*	*2.2.79*	*1.3.79*
	CYANOPHYTA					
1.	*Chroococus minor*	–	++	+	+	+
2.	*C. mintus*	+	+	+	+	+
3.	*Lyngbya birgie*	–	++	+	+	+
4.	*Merismopedia minima*	+	+	+	+	+
5.	*M. punctata*	–	–	+	+	+
6.	*Microcystis firma*	++	++	++	++	++
7.	*Nodularia spumigena*	–	–	–	–	+
8.	*Oscillatoria amphibia*	+	++	+	+	+
9.	*O. limosa*	–	+	+	+	+
10.	*Rhabdoderma minima*	–	–	–	+++	+++
	BACILLARIOPHYTA					
11.	*Fragilaria construens*	–	+	+	+	+
12.	*Navicula halophila*	+	++	++	++	+++
13.	*Nitzschia obtusa var scalpelliformis*	+	++	++	++	+++
	XANTHOPHYTA					
14.	*Heterochloris sp.*	–	–	–	–	+

EUGLENOPHYTA					
15. *Euglena rustica*	+	++	+	+	+
16. *Phacus arnoldi*	–	–	–	–	–
17. *P. gigas*	+	+	++	++	++
18. *Trachelomonas volvocina*	–	+	+	++	++
CHLOROPHYTA					
19. *Spondylmorum quaternarium*	–	–	–	–	++
20. *Ankistrodesmus convolutus*	+	+	++	++	++
21. *Coelastrum astroideum*	+	+	+	+	+
22. *Dictyosparium pulchellum*	–	–	+	+	+
23. *Pediastrum tetras*	+	+	+	++	++
24. *Scenedesmus acutiformis*	–	+	+	+	+
25. *S. arcuatus* var. *capitatus*	+	+	+	++	++
26. *S. bijuga*	++	++	++	++	++
27. *S. communis* Hegew. (= S. quadricauda)	+	+	+	+	+
28. *S. obtusus*	–	–	–	+	+
29. *S. opoliensis*	+	+	+	+	+
30. *Closterium acerosum*	–	–	+	++	+
31. *Cosmarium regnellii*	+	++	+++	+++	++
32. *Staurastrum muticum* var. *ellipticum*	+	++	+++	++	++

++++ = Dominant; +++ = Abundant; ++ = Common; + = rare.

The water was covered by an algal bloom during the entire period of study. The pH of the water was high almost throughout. This was correlated well with the absence of free CO_2 and presence of carbonates and bicarbonates in complementary concentrations. The N/P ratio was around 2.0 till January and around 1.0 in February and March. The other parameters were more or less constant (Table 17.1).

This tank was characterised by a persistant bloom of *Microcystis aeruginosa* and hence a very low species diversity. Besides *M. aeruginosa* with *Phormidium mucicola* occurring in its colonial mucilage and *Nitzschia obtusa* which was a co-dominant species, the phytoplankton consisted of a few Chlorococcalean and Cyanophycean members with one or two stray Euglenophytes (Table 17.3).

During the period of rains in October and November 1978 *Arthrospira massartii* occurred in large abundance along with *A. jennerri* and *Eudorina elegans*. This was well correlated with the higher amounts of bicarbonates. Their disappearance during the ensuing dry period was similarly correlated with the lower bicarbonate levels and presence of carbonates (Tables 17.1 and 17.3).

3. Parthasarathy Temple Tank

This tank is also in the heart of the city but close to the beach. It measures 96 metres × 64 metres with only about 2 metres deep water at the time of study and has access on all four sides. It is also bereft of tree cover.

Small patches of species of *Lemna, Nymphaea* and Grasses were found from February to April and *Pistia* sp. occurred in April only. Water was neutral to somewhat alkaline with some free CO_2 and moderate amounts of bicarbonate. The N/P ratio was 2.2 in February sample unlike the other two tanks. However, it dropped to less than 1.0 later. All other parameters did not fluctuate markedly (Table 17.1).

The phytoplankton flora was highly diverse (Table 17.4) with a total number of 54 species, the highest recorded among all the tanks studied. Of these, nearly 60% was represented by Chlorococcalean members with the rest constituted by Cyanophyceae, Euglenophyceae, Xanthophyceae and diatoms in the stated order of abundance. Only a single desmid was observed whereas unicellular Xanthophyceae and Euglenophyceae together constituted about 20% of the flora.

In this tank samples were taken long after the monsoon rains, by which time a large number of taxa had developed. As there was no shade nor much of blooms (only patchy) to compete for nutrients and sunlight,

Table 17.3 : Changes in Phytoplankton Community in Vadapalani Temple Tank

	Phytoplankton	Date of Sampling				
		23.11.78	3.1.79	18.1.79	1.2.79	1.3.79
	CYANOPHYTA					
1.	*Arthrospira massartii*	+++	–	–	–	–
2.	*A. jenneri*	+	–	–	–	–
3.	*Microcystis aeruginosa*	Bloom	Bloom	Bloom	Bloom	Bloom
4.	*Oscillatoria amphibia*	+	+	+	+	+
5.	*Phormidium mucicola*	++++	++++	++++	++++	++++
6.	*Spirulina substilissima*	–	–	–	+	–
7.	*S. meneqhiniana*	–	–	–	++	–
8.	*Spirulina* sp.	+	+	+	+	–
	BACILLARIOPHYTA					
9.	*Gomphonema parvulum*	–	–	+	–	–
10.	*Nitzschia obtusa* var. *scalpelliformis*	+++	+++	++++	+++	–
	EUGLENOPHYTA					
11.	*Euglena* sp.	+	+	+	–	–
12.	*Euglena* sp. (encysted condition)	–	–	–	+	+++
13.	*Lepocinclis teres f. parvula*	–	–	–		++

Table 17.3 (Contd.)

	Phytoplankton	Date of Sampling				
		23.11.78	3.1.79	18.1.79	1.2.79	1.3.79
	CHLOROPHYTA					
14.	*Eudorina elegans*	+	–	–	–	+
15.	*Ankistrodesmus falcatus*	+	+	+	+	–
16.	*Coelastrum microporam*	+	+	+	+	+
17.	*Micractinium pusillum*	–	–	–	+	+
18.	*Oocystis* sp.	–	–	–	+	+
19.	*Scenedesmus armatus* var. *bicaudatus*	+++	+++	+++	++	++
20.	*S. bijuga*	+	+	+	+	+

++++ = Dorminant; +++ = Abundant; ++ = Common; + = rare.

Table 17.4 : Changes in Phytoplankton Community in Parthasarathy Temple Tank

Phytoplankton	Date of Sampling			
	16.2.79	1.3.79	17.3.79	2.479
CYANOPHYTA				
1. *Anabaenopsis arnoldii*	–	–	–	+
2. *Chroococcus minor*	+	+	+	+
3. *C. turgidus*	–	–	–	+
4. *Coelosphaerium kuetzingianum*	–	–	–	+
5. *Merismopedia minima*	++++	++++	++++	++++
6. *M. tenuissima*	+	+	+	+
7. *Oscillatoria amphibia*	+	+	+	+
BACILLARIOPHYTA				
8. *Cyclotella* sp.	–	–	–	+
9. *Navicula halophila*	–	–	–	+
10. *Nitzschia obtusa* var. *scalpelliformis*	+	+	+	+

Table 17.4 (Contd.)

	Phytoplankton	Date of Sampling			
		16.2.79	1.3.79	17.3.79	2.4.79
	XANTHOPHYTA				
11.	*Goniochloris allorgei*	–	–	–	++
12.	*G. fallax*	+	+	+	++
13.	*G. mutica*	+++	+++	+++	+++
14.	*Pleurogaster lunaris*	–	–	–	+
15.	*Pseudostaurastrum hastatum*	+	++	++	++
	EUGLENOPHYTA				
16.	*Euglena oxyurix*	+	+	+	+
17.	*Euglena rustica*	++	++	++	++
18.	*Phacus acuminatus* var. *granulata*	++	++	++	++
19.	*P. tortus*	–	–	+	+
20.	*Trachelomonas volvocina*	++++	+++	+++	++++
	CHLOROPHYTA				
21.	*Ankistrodesmus arcuatus*	–	–	–	+
22.	*A. convolutus*	+++	+++	+++	+++
23.	*A. falcatus* var. *articularis*	-	++	+	+
24.	*A. falcatus* var. *radiatus*	-	-	-	+
25.	*Coleastrum astroideum*	-	-	-	+

26.	*C. microporum*	++	++	++	++
27.	*Crucigenia crucifera*	+++	+++	++	++
28.	*C. quadrata*	+	+	+	+
29.	*C. tetrapedia*	++	++	+	+
30.	*Dictyosphaerium pulchellum*	++	++	++	++
31.	*Didymocystis bicellularis*	+	+	+	+
32.	*Kirchneriella contorta*	+	+	+	++
33.	*K. irregularis*	++	++	++	++
34.	*L. lunaris*	++	++	++	++
35.	*Oocystis solitaria*	++	+	+	++
36.	*Oocystis* sp.	+	+	+	+
37.	*Pediastrum duplex*	++	++	++	++
38.	*Scenedesmus armatus*	+	+	+	+
39.	*S. armatus* var. *bicaudatus*	+	+	+	+
40.	*S. bijuga*	++	++	++	++
41.	*S. dimorphus*	+++	+++	+++	+++
42.	*S. dimorphus f. tortus*	++	++	++	++
43.	*S indicus*	-	-	-	+
44.	*S. obliqis*	+	+	+	+
45.	*S. obtusus*	+	+	+	+
46.	*S. opoliensis*	+	+	+	++
47.	*S. platydiscus*	+	+	+	++
48.	*Siderocelis elegans*	-	-	-	+

(Contd.)

Table 17.4 (Contd.)

	Phytoplankton	Date of Sampling			
		16.2.79	1.3.79	17.3.79	2.4.79
49.	*Tetraedron minimum*	+	++	++	++
50.	*Tetrastrum heteracanthum*	-	-	-	+
51.	*T. heteracanthum f. elegans*	-	-	-	+
52.	*T. punctatum*	-	-	-	+
53.	*T. staurogeniaeforme*	-	-	-	+
54.	*Cosmorium* sp.	-	-	-	+

++++ = Dorminant; +++ = Abundant; ++ = Common; + = rare.

a good and increasingly diverse phytoplankton community developed (Table 17.4). The relatively low water depth probably facilitated a quick turnover of nutrients through decomposition aided by the higher temperature and mixing.

Discussion

The three tanks studied were distinctly different in their phycobiocoenoses although they were situated within a radius of about 15 kms. Similar observations were made in earlier studies on coastal beach ponds (Jeeji Bai, 1990) and other localities (Zafar, 1964; Sreenivasan, 1968; Chatterjee and Mohanty, 1990). Gangadhareswarar and Parthasarathy tanks were similar in the low biomass and high species diversity of their phytoplankton flora. In both tanks Chlorococcalean and Cyanophycean species predominated the floras with Eugelnophytes following next in abundance (Tables 17.2 and 17.4). This general similarity in their phycobiocoenoses was reflected also in their similar water chemistry (Table 17.1) and the macrophytic vegetation occurring in them (*Lemna, Nymphaea* sp.). However, the individual species constituting the biocoenoses of these two tanks were quite different except for a few common species (Tables 17.2 and 17.4).

Chlorococcalean predominance in phytoplankton associations were observed in different types of waters, (Anand 1975; Kant and Kachro, 1977; Hosmani and Bharati, 1980; Somashekar 1984; and Sengar and Sharma, 1985). However, co-dominance of desmids, diatoms and sometimes Euglenophytes as reported in many of these studies was not observed in the present study. Zafar (1964; 1967) in his studies on fish ponds of Hyderabad pointed out that Pearsall's basic ratio showed a better correlation with the occurrence of Chlorococcales, desmids and diatoms; values of 0.02-1.2 favouring diatoms, of 2.0-3.5 favouring Chlorococcales and desmids, values above 3.5 not further influencing their occurrence. As these parameters were not analysed in this study, such a correlation could not be determined. The development of a fair number of Xanthrophycean and Chlorococcalean nannoplankton in Parthasarathy tank (Table 17.4) is probably due to the greater light and nutrient availability besides the small pore size of the plankton net used.

The Vadapalani tank was characterised by a more or less stable bloom of *Microcystis aeruginosa,* hence a high biomass and low species diversity (Table 17.3). Bloom of *M. aeruginosa* reported earlier in Gangadharareswarar tank by Ganpati *et al.*, (!953) was replaced by an entirely different flora in this study (Table 17.2).

Blooms of *M. aeruginosa* were reported frequently in lakes and ponds from all over India under widely different climatic and hydrological conditions. It formed either a permanent bloom throughout the year (Ganpati *et al.*, 1953; Sreenivasan *et al.*, 1973) or often occurring in close association with monsoon rains (Grover *et al.*, 1978; Singh, 1968; Singh 1981). Bloom formation was correlated with low oxygen and phosphate levels (Khan and Seenayya, 1985), intermediate Composite Trophy Index values (Hegde and Bharati, 1984) and high oxygen levels (Hosmani and Bharati, 1980) along with high albuminoid nitrogen (Jayangoudar, 1964). It appears from this study that pH values of 9.0 and above and high bicarbonate/carbonate alkalinities favour *Microocystis* blooms as also reported by Sreenivasan (1968).

The obvious difference in the phycocoenoses of the three tanks could be correlated more or less directly with the presence or absence of free CO_2 and carbonates and therefore, with the pH value of the water. The other significant difference was in the ratio of nitrates and phosphates which probably exerted an influence on the other differences in their floristic composition.

Conclusions

The general view based on studies of large lakes that oligotrophic waters have a flora with high species diversity cannot be extrapolated to small water bodies of the tropics. In small pond ecosystems of the tropics such as those studied here, the degree of species diversity is not determined by the low nutrient status (mainly phosphates) alone but by a variety of other conditions which are brought about by monsoon rains. High bicarbonate alkalinities, and pH of the water generally favour blue-green algal blooms. It is not absolute amounts of phosphate alone but a complexity resulting from an interaction of both physical and chemical factors which determine the quality and quantity of phytoplankton of these waters.

An attempt to compare with other similar studies showed that no two studies were in total agreement with regard to physico-chemical factors influencing the occurrence or abundance of the different groups. It appears that a multivariant analysis may be necessary to find certain reliable and generally applicable criteria in such studies. The coexistence of widely different groups may also be due to the existence of favourable "niches" within the larger pond habitat. Finally, as pointed out by Wetzel (1979) and in an earlier study (Jeeji Bai, 1990) the influence of littoral and floating macrophytic vegetation cannot be undermined in such small pond ecosystems.

Acknowledgements

We thank the Director, Centre of Advanced Study in Botany, University of Madras and the University Grants Commission for the award of a teacher fellowship to D. Lakshmi, during the tenure of which this investigation was carried out for the M. Phil. Degree.

References

APHA, AWWA & WPCF, (1962) *Standard Methods for the Examination of Water and Wastewater Including Bottom Sediments and Sludges*, New York, pp. 874.

Anand, V.K., (1975) A checklist of planktonic algae from Mansar lake, Jammu. *Phykos,* V. 14, pp. 77-80.

Barnes, H., (1959) Apparatus and Methods of Oceanography. Part I Chemical, George Allen & Unwin Ltd. London. 341 pp.

Chacko, P.I., (1972) Blue Green algae of fish ponds in Tamil Nadu, India. In : T.V. Desikachary (ed.) *Taxonomy and Biology of Blue Green Algae*, Univ. of Madras, pp. 448-54.

Chatterjee, A.A.K. and Mohanty, R.C., (1990) Distributional patterns of phytoplankton in two freshwater lakes of Puri district. In: Raja Rao, V.N. (Ed.) *Perspectives in Phycology*, Today and Tomorrow Print & Publ. New Delhi, pp. 171-81.

Ganapati, S.V., (1941). Studies on the chemistry and biology of ponds in the Madras city (a) Seasonal changes in the physical and chemical conditions of a garden pond containing abundant aquatic vegetation, *J. Madras Univ*. V. 13, pp. 55-69.

Ganpati, S.V., Chacko, P.I. and Srinivasan, A., (1953) Hydrobiological conditions of the Gangadhareswar temple tank, Madras, *J. Asiatic Soc. Sci.* V. 19 (2), pp. 149-59.

Grover, I.S., Kalra, S.S. and Puri, S., (1978) Hydrobiological studies of two fish ponds of Ludhiana, *Phykos,* V. 17, pp. 39-48.

Hegde, G.R. and Bharati, S.G., (1984) Ecological studies in ponds and lakes of Dharwad, Trophic status. *Phykos.* V. 23, pp. 71-74.

Hosmani, S.P. and Bharati, S.G., (1980) Limnological studies in ponds and lakes of Dharwar. Comparative phytoplankton ecology of four water bodies. *Phykos*. V. 19, pp. 27-43.

Iyengar, M.O.P., (1920) Observations on the Volvocaceae of Madras. *J. Indian Bot. Soc.* V. 1, pp. 330-36.

Iyengar, M.O.P., (1939). Algal problems peculiar to the tropics with special reference to India. In. *Proc. 25th Indian Sci. Congr.* 4, pp. 1-8.

Jayangoudar, I.S., (1964) A bio-ecological study of Nuggikari lake in Dharwar, Mysore state, India, *Hydrobiologia,* V. 29, pp. 156-64.

Jeeji Bai, N., (1990) A holistic approach to the study of small pond ecosystems of South India In : Raja Rao, V.N. (Ed.) *Perspectives in Phycology*; Today and Tomorrow Print. & Publ. New Delhi, pp. 201-07.

Kant, S. and Kachroo, P., (1977). Limnological studies in Kashmir lakes 1: Hydrological features, composition and periodicity of phytoplankton in the Dal and Nagin Lakes *Phykos*, V. 16 (1-2), pp. 77-97.

Khan, M.M. and Seenayya, G., (1985) Ecology of planktonic blue green algae in the profiles of industrially polluted Husain Sagar Lake, Hyderabad, India. *Phykos*. V. 24, pp. 140-48.

Sengar, R.M.S. and Sharma, K.D., (1985). Quantitative study of phytoplankton of Yamuna river, *Phykos*, V. 24, pp. 117-24.

Somashekar, R.K., (1984) Contribution to the algal flora of river Kapila, Karnataka. I. Cyanophyceae and Chlorophyceae *Phykos*. V. 23, pp. 116-24.

Singh, M., (1968). Phytoplankton periodicity in a small lake near Delhi. 2. Seasonal fluctuations of the main groups of phytoplankton. *Phykos,* V. 7, pp. 126-35.

Singh, S.R., (1981) Studies on seasonal composition and horizontal distribution of phytoplankton of Suraha lake. *Phykos*, V. 20, pp. 34-43.

Sreenivasan, A., (1968) Limnology of tropical impoundments V. studies of two upland impoundments in Nilgris, Madras State, (India). *Phykos*. V. 7(1 & 2), pp. 144-60.

Sreenivasan, A. Soundararaj, R. and Franklin, T., (1973) Diurnal and seasonal changes in a productive shallow tropical pond. *Phykos*, V. 12 (1 & 2), pp. 86-103.

Theroux, R., Eldridge, E.F. and Le Roy, M., (1943) *Laboratory Manual for Chemical and Bacterial Analysis of Water and Sewage.* McGraw Hill, Lond. pp. 274.

Thomasson, K., (1979). Heleoplankton from a pool in South Tondelag province, Central Norway, *Gunnera,* V. 32, pp. 1-23.

Wetzel, R.G., (1979) The role of the littoral zone and detritus in lake metabolism. In Rhode, W., Likens, G.E., and Serruya, C., (eds.) *Lake*

Metabolism and Management, Arch. Hydrobiol. Beih. Ergegn. Limnol. V. 13, pp. 145-62.

Zafar, A.R., (1964) On the ecology of algae in certain fish ponds of Hyderabad, India II. Distribution of unicellular and colonial forms. *Hydrobiologia*, V. 24, p. 556-66.

Zafar, A.R., (1967) On the ecology of algae in certain fish ponds of Hyderabad, India. III. The periodicity. *Hydrobiologia*, V. 30, pp. 96-112.

18

Physico-Chemical and Biological Characteristics of Balal Lake, Bodhan, Nizamabad District, Andhra Pradesh

A. Sayeeswari
and
M.A. Singara Charya

Introduction

Owing to the indescriminate exploitation of freshwater resources of India, the limnological studies of such water bodies are being considered ecologically. Environmentally, these aquatic bodies are facing severe threat because of rapid urbanisation and industrialisation.

There is some information on the limnology of North Indian tropical and sub-tropical lakes (Swarup and Singh, 1979; Iqbal, 1986; Nair and Pandaya, 1986; Rao and Durve, 1989; Vijay and Bhatnagar, 1989; Moore and Gajjar, 1990). Such information on the lakes of South India is confined to the works of Ganapati (1941), Sreenivasan (1965), Munavar (1970), Hosmani and Bharati (1980) and Seshavatharam (1982). But there is comparatively less information on the freshwater bodies of Andhra Pradesh and no such studies were undertaken in Nizamabad district of Teleangan region. Hence, a hitherto uninvestigated Balal lake was selected for limnological studies for a period of one year (1989-90).

The present study has been aimed at assessing the physico-chemical and biological characteristics of the lake which has a capacity of 346.92 MC ft and catchment area of 120.43 sq km stretching between longitude 77°55′, E, latitude 18°37′N. The lake is situated about 2.5 km to Bodhan town. The Nizam Sugar Factory (NSF) is 3 km down to this lake. The Nizam Sagar main canal passes at about 14 km away from this lake. The irrigated land under this canal is the main catchment area of the lake (Map 1). There are three outlets from the lake : (i) Bommanna sluice, (ii) Papanagamma sluice, and (iii) NSF feeder channel. Of these, the first two are meant for irrigation (about 1079.3 ha in kharif and about 405 ha in rabi season). The NSF feeder channel is meant for the needs of Nizam Sugar Factory.

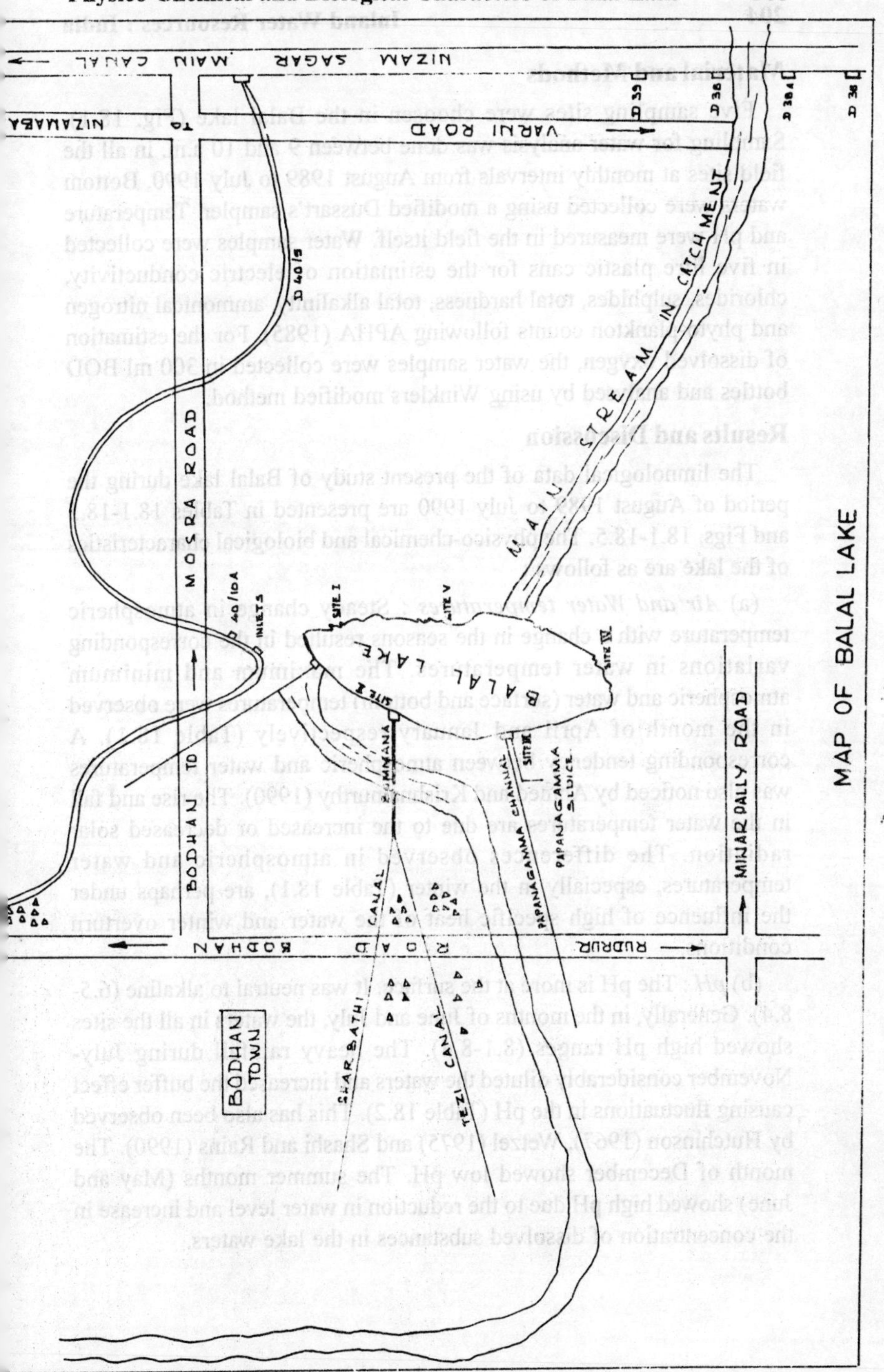
MAP OF BALAL LAKE
NIZAM SAGAR MAIN CANAL
VARNI ROAD
TO NIZAMABA
BODHAN TO MOSRA ROAD
MANU STREAM IN CATCHMENT
BALAL LAKE
INLETS
PAPANGAMMA SLUICE
MINAR DALLY ROAD
BODHAN TOWN
TEZI CANAL
BODHAN RUDRUR ROAD

Material and Methods

Five sampling sites were choosen in the Balal lake (Fig. 18.1). Sampling for water analysis was done between 9 and 10 a.m. in all the field sites at monthly intervals from August 1989 to July 1990. Bottom waters were collected using a modified Dussart's sampler. Temperature and pH were measured in the field itself. Water samples were collected in five litre plastic cans for the estimation of electric conductivity, chlorides, sulphides, total hardness, total alkalinity, ammonical nitrogen and phyto-plankton counts following APHA (1985). For the estimation of dissolved oxygen, the water samples were collected in 300 ml BOD bottles and analysed by using Winklers modified method.

Results and Discussion

The limnological data of the present study of Balal lake during the period of August 1989 to July 1990 are presented in Tables 18.1-18.5 and Figs. 18.1-18.5. The physico-chemical and biological characteristics of the lake are as follows :

(a) *Air and Water temperatures* : Steady change in atmospheric temperature with a change in the seasons resulted in the corresponding variations in water temperatures. The maximum and minimum atmospheric and water (surface and bottom) temperatures were observed in the month of April and January respectively (Table 18.1). A corresponding tendency between atmospheric and water temperatures was also noticed by Ahmed and Krishnamurthy (1990). The rise and fall in the water temperatures are due to the increased or decreased solar radiation. The differences observed in atmospheric and water temperatures, especially in the winter (Table 18.1), are perhaps under the influence of high specific heat of the water and winter overturn conditions.

(b) *pH* : The pH is more at the surface. It was neutral to alkaline (6.5-8.4). Generally, in the months of June and July, the waters in all the sites showed high pH ranges (8.1-8.4). The heavy rainfall during July-November considerably diluted the waters and increased the buffer effect causing fluctuations in the pH (Table 18.2). This has also been observed by Hutchinson (1967), Wetzel (1975) and Shashi and Rains (1990). The month of December showed low pH. The summer months (May and June) showed high pH due to the reduction in water level and increase in the concentration of dissolved substances in the lake waters.

Table 18.1 : Air and water temperatures of Balal Lake.

Month & Year	*Site I*	*Site II*	*Site III*	*Site IV*	*Site V*
August 1989	28.1[a]/27.4[b]/25.2[c]	28.0/27.6/25.4	29.5/27.8/26.2	29.0/28.3/26.4	28.5/27.2/25.6
September '89	31.5/29.0/28.6	32.0/29.5/28.1	32.0/30.2/27.3	32.5/30.1/28.4	31.5/29.5/28.4
October '89	32.0/31.0/27.6	33.1/31.3/26.5	33.5/29.6/27.1	32.3/28.6/26.8	32.2/28.5/25.6
November '89	31.0/29.5/27.5	31.5/29.2/26.2	31.2/29.5/27.2	30.4/29.5/27.2	30.4/30.2/28.5
December '89	29.5/26.0/24.5	28.4/26.5/24.4	28.5/24.5/23.1	29.0/26.2/24.7	28.4/24.6/21.7
January '90	27.0/24.0/20.4	28.1/25.1/23.1	28.5/24.2/23.3	27.4/22.6/20.8	27.3/22.3/20.7
February '90	29.0/26.2/24.4	29.5/27.5/25.2	30.2/29.3/27.5	29.1/27.3/26.1	29.5/26.5/24.4
March '90	33.3/29.0/27.5	34.3/31.2/28.5	33.2/33.0/29.4	34.1/30.8/29.5	33.2/29.6/28.3
April '90	36.0/32.0/31.3	36.5/33.0/31.4	35.6/32.8/30.3	36.2/31.7/29.8	35.4/32.7/29.2
May '90	28.1/26.0/24.7	27.3/25.2/22.3	28.2/26.3/24.6	28.4/26.8/23.7	28.2/26.2/24.6
June '90	32.8/32.1/31.3	32.9/32.0/31.6	33.1/31.9/29.1	32.3/31.7/30.1	32.4/31.4/28.7
July '90	29.3/29.0/28.2	29.2/28.6/26.7	29.4/28.2/27.0	30.5/29.1/27.5	29.2/28.7/28.1

a – Air temperature (in °C)
b & c – Water temperatures (in °C) surface and bottom respectively.

Table 18.2 : pH of waters at the surface and bottom levels of Balal Lake.

Month & Year	*Site I*	*Site II*	*Site III*	*Site IV*	*Site V*
August 1989	7.5*/7.4**	7.8/7.6	8.0/7.9	7.6/7.4	7.5/7.5
September '89	8.2/8.0	7.5/7.4	7.8/7.6	7.9/7.7	7.8/7.6
October '89	7.8/7.9	8.2/8.0	7.8/7.6	8.2/8.1	7.5/7.5
November '89	7.5/7.4	7.3/7.2	7.6/7.3	7.4/7.4	7.7/7.6
December '89	7.4/6.8	7.1/7.1	7.5/7.3	7.3/7.2	7.9/7.9
January '89	7.8/7.0	7.9/7.1	8.2/8.0	7.3/7.2	8.1/8.0
February '90	7.8/7.0	7.6/7.2	8.1/8.1	8.2/8.1	7.8/7.5
March '90	7.8/7.8	7.9/7.2	8.3/7.3	7.8/7.6	7.6/7.5
April '90	7.6/7.5	7.2/7.1	7.6/7.1	8.1/7.8	8.2/8.1
May '90	8.0/7.8	7.3/7.3	7.8/6.5	7.8/6.9	7.8/7.6
June '90	8.2/7.8	8.3/7.6	8.1/7.5	8.4/7.8	8.1/8.0
July '90	8.2/7.6	8.3/8.0	8.0/7.8	8.2/7.8	7.8/7.7

* surface level
** bottom level

(c) *Electric conductivity* : High electric conductivity in the lake waters was observed in all the sites in November 1989 (Table 18.3), while minimum conductivity was recorded in June for the sites III-V, May and July for site II and January for site I.

Table 18.3 : Electric Conductivity (milli mhos/cm) of Balai Lake.

Month & Year	*Site I*	*Site II*	*Site III*	*Site IV*	*Site V*
August 1989	1.06	1.08	1.02	1.12	1.00
September '89	1.21	1.24	1.12	1.00	1.08
October '89	1.32	1.31	1.44	1.04	1.37
November '89	1.48	1.52	1.56	1.51	1.62
December '89	1.34	1.32	1.40	1.38	1.50
January '90	0.98	0.96	0.85	1.10	1.20
February '90	1.31	1.26	0.92	1.30	1.21
March '90	0.84	1.20	1.10	0.75	0.98
April '90	1.10	1.30	1.20	0.90	1.10
May '90	1.20	0.90	1.30	1.10	1.40
June '90	1.20	1.10	0.80	0.60	0.80
July '90	1.30	0.90	1.00	1.00	1.20

* surface level
** bottom level

(d) *Dissolved oxygen* : The dissolved oxygen (DO) at the surface and bottom levels in selected sites of Balal lake was recorded in Table 18.4. The maximum amount of DO estimated in surface waters was in the month of July for sites I-IV and June for site V. The minimum DO estimated was in February. In the bottom waters too the maximum DO was noticed in the months of June and July, while the minimum DO varied from site to site spreading over the year. The reason for the maximum production of DO is because of the presence of photosynthetic organisms. The minimum and maximum concentrations of DO in the waters of five sites of the lake are correlated to minimum and maximum phytoplankton populations. Such observations were also made by Schindler (1971), Lande (1973) and Saad (1973).

(e) *Chlorides* : Slight variations were observed in chloride concentrations at different sites (Fig. 18.1). Chloride concentration was maximum in all the sites (I-V) in June. The heavy rainfall during this period might have supplemented its source from the catchment area (Hutchinson, 1967). It was minimum during monsoon and winter seasons. The works of Cole (1975), Unni (1984) and Shashi and Raina (1990) are in accordance with the present results.

(f) *Sulphides* : These were present in all the sites (Fig. 18.2) throughout the year but in low concentrations. The minimum (0.80 ppm) and maximum (1.46 ppm) sulphide concentrations are in site V in the months

Table 18.4 : Dissolved Oxygen (in ppm) at the surface and bottom levels of Balal Lake.

Month & Year	*Site I*	*Site II*	*Site III*	*Site IV*	*Site V*
August 1989	8.5*/4.3*	7.8/2.7	7.4/3.2	8.6/8.2	8.6/6.4
September '89	7.6/3.6	7.4/1.2	7.5/2.9	7.2/3.6	7.4/1.3
October '89	10.2/6.4	4.6/2.5	5.9/3.8	5.5/2.6	6.9/3.3
November '89	7.4/4.6	7.9/4.1	8.8/4.3	8.3/3.9	8.5/4.5
December '89	6.0/4.8	8.5/3.8	8.3/4.0	7.6/4.1	9.4/3.2
January '90	4.6/3.9	7.4/4.1	7.9/4.4	7.6/3.8	7.5/4.0
February '90	5.3/3.4	6.8/4.2	6.4/3.1	4.6/2.9	6.8/3.6
March '90	8.9/3.2	7.8/4.5	8.2/3.5	8.2/3.4	8.4/4.6
April '90	9.3/3.6	7.6/6.0	7.8/4.8	7.8/2.7	8.2/3.7
May '90	8.6/3.5	9.4/6.0	8.8/4.8	8.4/7.3	7.3/6.8
June '90	9.4/7.0	8.2/6.1	8.1/7.6	5.8/5.5	9.4/8.2
July '90	10.2/4.1	9.7/7.3	10.2/8.8	8.4/9.7	7.3/7.3

* surface level

** bottom level

Table 18.5 : Mean monthly Phytoplankton populations (cells/L, I * 10^3) of Balal Lake.

Month & Year	Site I					Site II					Site III					Site IV					Site V				
	A	*B*	*C*	*D*	*E*	*A*	*B*	*C*	*D*	*E*	*A*	*B*	*C*	*D*	*E*	*A*	*B*	*C*	*D*	*E*	*A*	*B*	*C*	*D*	*E*
August 1989	45	18	12	1	76	42	18	12	1	73	46	26	18	–	90	40	14	25	1	80	32	20	18	2	72
September '89	24	34	14	2	74	28	24	36	–	88	32	46	22	–	100	42	16	10	3	71	36	19	11	9	75
October '89	10	26	9	–	45	29	46	18	1	94	16	8	9	2	35	24	12	10	2	48	10	24	16	3	53
November '89	28	24	18	3	73	16	52	34	2	104	14	19	24	–	57	24	14	22	4	64	14	28	22	2	66
December '89	16	34	10	3	63	15	44	26	2	47	12	42	18	1	73	22	23	18	3	66	6	42	33	4	85
January '90	14	38	24	–	76	14	26	12	3	55	4	38	34	3	79	12	18	12	2	44	18	21	34	2	75
February '90	26	42	6	1	75	24	62	28	1	115	42	22	8	4	76	14	23	14	2	56	24	14	30	3	71
March '90	43	84	4	2	133	46	54	24	3	127	38	84	18	1	141	42	44	8	–	94	26	46	22	1	95
April '90	32	92	10	20	136	24	126	16	8	174	26	48	14	1	89	28	12	15	3	58	52	38	9	4	103
May '90	26	96	12	1	135	28	124	16	2	170	12	14	13	2	41	19	5	6	1	31	22	20	8	1	51
June '90	24	62	14	3	103	13	78	14	1	106	10	64	9	3	86	7	2	2	–	11	21	16	4	2	43
July '90	23	24	18	4	69	48	82	13	2	145	30	32	16	2	80	52	14	14	2	82	26	34	7	5	72

A – Chlorophyceae
B – Myxophyceae
C – Bacillariophyceae
D – Others
E – Total

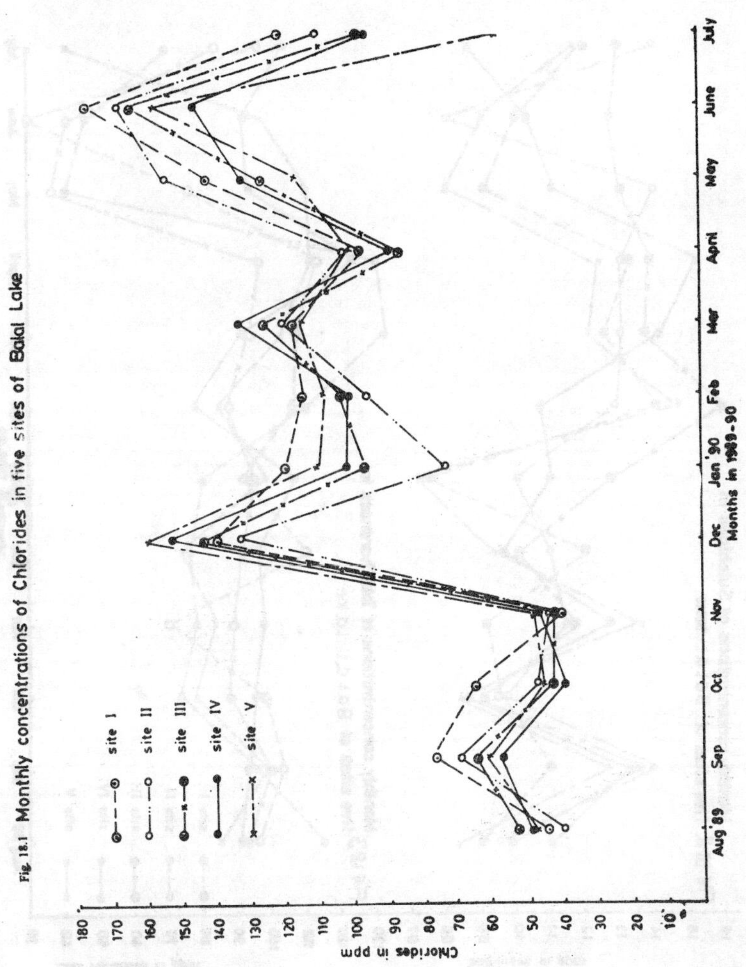

Fig. 18.1 Monthly concentrations of Chlorides in five sites of Balal Lake

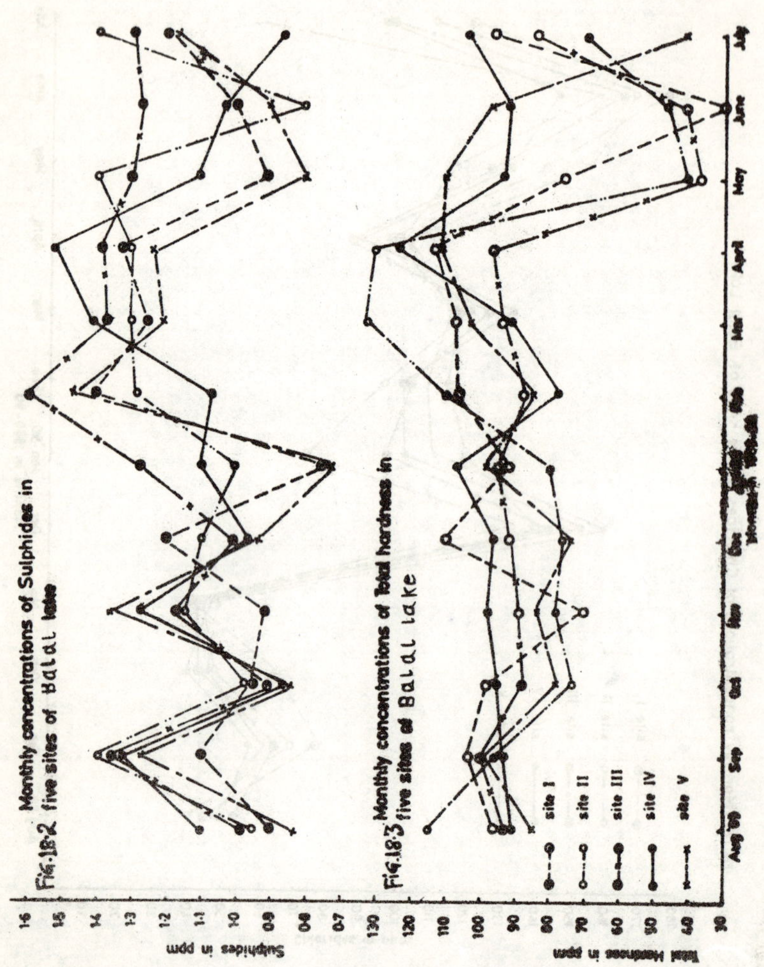

Fig.18.2 Monthly concentrations of Sulphides in five sites of Balal lake

Fig.18.3 Monthly concentrations of Total hardness in five sites of Balal lake

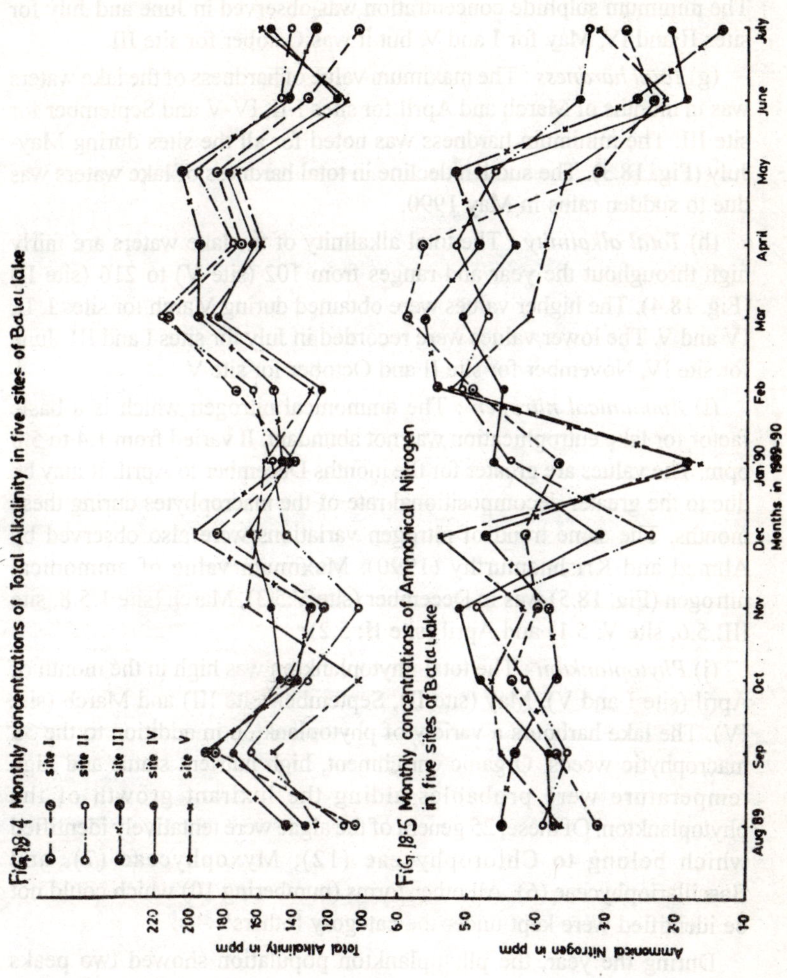

Fig. 18.4 Monthly concentrations of Total Alkalinity in five sites of Balal lake

Fig. 18.5 Monthly concentrations of Ammonical Nitrogen in five sites of Balal Lake

of May and February, respectively. The maximum amount of sulphides were in summer months (March-May), except in site II (September). The minimum sulphide concentration was observed in June and July for sites II and IV, May for I and V, but it was October for site III.

(g) *Total hardness* : The maximum value of hardness of the lake waters was in months of March and April for sites I-II, IV-V and September for site III. The minimum hardness was noted for all the sites during May-July (Fig. 18.3). The sudden decline in total hardness of lake waters was due to sudden rains in May 1990.

(h) *Total alkalinity* : The total alkalinity of the lake waters are fairly high throughout the year and ranges from 102 (site V) to 216 (site II) (Fig. 18.4). The higher values were obtained during March for sites I, II, IV and V. The lower values were recorded in July for sites I and III, June for site IV, November for site II and October for site V.

(i) *Ammonical nitrogen* : The ammonical nitrogen which is a basic factor for lake eutrophication was not abundant. It varied from 1.4 to 5.8 ppm. The values are greater for the months December to April. It may be due to the greater decompositional rate of the macrophytes during these months. The same trend of nitrogen variations were also observed by Ahmed and Krishnamurthy (1990). Maximum value of ammonical nitrogen (Fig. 18.5) was in December (site V:5.3), March (site 1:5.8, site III: 5.6, site V: 5.1) and April (site II: 5.2).

(j) *Phytoplankton* : The total phytoplankton was high in the month of April (site I and V), May (site II), September (site III) and March (site IV). The lake harbours a variety of phytoplankton in addition to the 33 macrophytic weeds. Organic enrichment, high nutrient status and high temperature were probables aiding the luxirant growth of the phytoplankton. Of these, 25 genera of the algae were tentatively identified which belong to Chlorophyceae (12), Myxophyceae (7), and Bascillariophyceae (6). All other forms (numbering 10) which could not be identified were kept under the category 'others'.

During the year, the phytoplankton population showed two peaks (bomodal : 'Summer peak' occurring between March and April and 'Winter Peak' occurring between August and September (Table 18.5).

Acknowledgements

Authors are thankful to Prof. Vidyavati, Head Department of Botany, Kakatiya University, Warangal for kind encouragement and providing necessary facilities. Our thanks are also to Dr. Vatsavaya S. Raju for useful discussion and to UGC, New Delhi for financial assistance.

References

Ahmed, M. and Krishnamurthy, R., (1990) Hydrobiological studies in Wohar reservoir, Aurangabad (Maharastra state), India J. Environ. Biol. V. 11, pp. 335-43.

Anonymous, (!985) '*Standard methods for the examination of water and wastewater*' 16th edition, American Public Health Association Publication, Washington, D.C., USA.

Cole, G.A., (1975) *A Text Book of Limnology*. C.V. Mosby Co. Saint Louis.

Ganapati, S.V., (1941) Studies on the chemistry and biology of ponds in the Madras city. Seasonal changes in the physical and chemical conditions of a garden pond containing aquatic vegetation. J. Madras Univ. V. 13, pp. 55-59.

Hosmani, S.P. and Bharati, S.G., (1980) Limnological studies in pond and lakes of Dharwad-Comparative phytoplankton ecology of four water bodies. Phykos, V. 19, pp. 27-43.

Hutchinson, G.E., (1967) *A Treatise on Limnology.* II. *Introduction to Lake Biology and Limnoplankton. John Wiley & Sons Inc.* New York.

Iqbal, A.K. (1986) A note on the limnology of the Baigul reservoir, U.P., India. J. Fisheries V. 33, pp. 119-23.

Lande, A., (1973) Primary production and other limnological features in an oligotrophic Norwegian lake. Hydrobiologia 41, pp. 335-44.

Moore, P.G. and Gajjar, H.J., (1990) Hydrobiological study of Palan pond, Valsad, Gujarat, J. Ecobiol. V. 2, pp. 128-35.

Munawar, M., (1970) Limnological studies on freshwater pond of Hyderabad, India-II. The biocenose distribution of unicellular and colonial phytoplankton in polluted and unpolluted environments. Hydrobiologia, V. 36, pp. 105-28.

Nair, R.K. and Pandya, S.C. (1986) A comparative ecological study of four freshmater bodies at Rajkot, North India. Intl. J. Ecol. Environ. Sci. V. 12, pp. 89-101.

Rao, N.G. and Durve, V.S., (1989) Cultural eutrophication of the lake Ranga Sagar, Udaipur, Rajasthan, India, J. Environ. Biol. V. 10, pp. 127-39.

Saad, M.A.H., (1973) Some limnological characteristics of the Nozh, a hydradrome near Alexandria, Egypt, Hydrobiologia 41, 477-500.

Schindler, D.W., (1971) Light, temperature and oxygen regimes of selected lakes in the experimental lakes area, North-Western Ontaria. J. Fish. Res. Canada V. 28, pp. 157-69.

Shashi, K. and Raina, A.K. (1990) Limnological studies of two ponds in Jammu-II. Physico-chemical properties. J. Environ. Biol. V. 11, pp. 137-44.

Seshavatharam, V., (1982) Limnological studies on Kondakarla lake-a freshwater body near Vishakhapatnam. Project Report No. 002/Biol/78.

Sreenivasan, A., (1965) Limnology of tropical impoundments-III. Limnology and productivity of Amaravati reservoir. Hydrobiologia 26, pp. 501-16.

Swarup, K. and Singh, S.R. (1979) Limnological studies of Suraha lake (Ballia)-I. Variations in the water quality. J. Indian Inland Fish. Soc. 11, pp. 22-32.

Unni, K.S. (1984) Limnology of sewage polluted tank in Central India. Intl. Revue Res. Hydrobiol. V. 69, pp. 553-65.

Vijay, V. and Bhatnagar, G.P. (1989) Primary productivity of phytoplankton in a eutrophic lower lake. Bhopal, India. Environ. Evol. V. 7, pp. 202-95.

Wetzel, R.G. (1975) *Limnology.* W.B. Saunders Company, Philadelphia.

19

Trend and Interannual Variability of Rainfall in Purulia district, West Bengal

T. N. TIWARI

AND

MILINA PRIYADARSHINI

Introduction

Water is one of the most precious gifts of nature. In the context of our burgeoning population, rapid industrialisation and growing demand on agriculture, there is an urgent need for utmost conservation and optimum utilization of this precious resource.

Since Independence, India has made great strides in the expansion of industrial base with the result that India is one of the ten most industrialised countries of the world today. Environmental degradation, however, is one of the most undesirable effects of poorly planned (or, worse still, unplanned) industrialisation. In addition to other well-known effects, environmental deterioration may disturb the well-established climatic patterns, and this may have far-reaching consequences on agriculture, forestry, etc.

There is some indication that the well-established rainfall patterns have already been disturbed in our country since floods and droughts appear to have become more frequent during the last 2-3 decades. Such disturbances are bound to have serious implications for our agricultural economy. It is necessary, therefore, to closely monitor such changes in the rainfall patterns due to large-scale industrialisation and other development activities.

In recent years the trends of rainfall patterns at Rourkela (Tiwari, 1989A; Tiwari and Patel, 1989; Patel and Tiwari, 1991) and Phulbani Town (Tiwari *et al.*, 1990) in Orissa and Sangli district in Maharashtra (Tiwari, 1989B) are studied by us. Continuing the investigations along similar lines, here it is reported on the trends of annual rainfall in Purulia district in West Bengal.

Rainfall in Purulia District

Purulia is the western most district in the State of West Bengal in the eastern part of India. Purulia has been classified as a drought-prone district by the Meteorological Department India. The area of Purulia district is about 6,529 km². This district has an undulating topography with rugged hilly terrain in its western and southern parts. The rivers Damodar and Subarnarekha flow through the northern and western boundaries of Purulia. In addition to these, three other rivers, viz., Kashi, Darakeswar and Silabati, flow through the district of Purulia and act as a natural drainage system.

Daily rainfall in Purulia district is recorded at the rainguage station located at Purulia town, set up by the Meteorological Department (New Delhi) in 1950. In addition to this, the Department of Agriculture, Government of West Bengal, had established 9 more rainguage stations in the 1970s. These stations, distributed throughout the district of Purulia, are located at the following towns: (1) Raghunathpur; (2) Kashipur; (3) Hura; (4) Man Bazar; (5) Bara Bazar; (6) Hatwara; (7) Bagmundi; (8) Jhaldah; and (9) Para. The available data on the annual rainfall recorded at these ten rainguage stations in Purulia district (Bhattacharjee and Chanda, 1988) over the 13-year period (1972-1984) are summarized in Table 19.1. This table gives the average value and range of the rainfall at the ten stations mentioned earlier for the period shown in the second column in each case. The last row of Table 19.1 shows data, based on the annual averages taken over the ten rainguage stations, for the whole of

Table 19.1 : Summary of annual rainfall data from Purulia District of West Bengal (Bhattacharjee and Chanda, 1988)

Station	*Period*	*Annual rainfall (cm)*	
		Average value	*Range*
Raghunathpur	1971-81	138.4	73.9–205.5
Kashipur	1972-84	116.6	69.0–161.8
Hura	1976-84	130.7	86.7–238.1
Man Bazar	1980-84	107.1	71.1–125.1
Purulia Town	1972-80	138.8	94.7–214.0
Bara Bazar	1979-84	111.4	67.6–196.3
Hatwara	1972-83	117.8	78.3–158.9
Bagmundi	1978-84	134.2	98.4–166.3
Jhaldah	1972-84	104.1	62.5–161.0
Para	1977-84	114.2	74.5-177.0
Purulia District	1972-84	122.4	88.5–194.4

Purulia district. Figure 19.1, on the other hand, shows average annual rainfall for the whole of Purulia district and its general trend for the period 1972-84.

Theoretical Considerations

The simplest method of studying the trend of annual rainfall is to treat it as a function of time and fit it to the best straight line. If the slope of such a straight line is positive, the rainfall is increasing with the passage of time. On the other hand, if the slope is negative, it indicates a declining trend of the annual rainfall.

Let y be a linear function of an independent variable x:

$$y = ax + b \quad \text{...(1)}$$

The parameters 'a' and 'b' in equation-1 are, respectively, the slope and intercept of the straight line defined by this equation. Let (x_i, y_i) be n pairs of the observed values of these variables (i = 1, 2....., n and n = no. of observations). Then according to the well-known method of least squares (Wonnacott and Wonnacott, 1981), the best value of the slope parameter 'a' is given by the relation

$$a = \frac{n\,\Sigma(x.y) - (\Sigma x) * (\Sigma y)}{n\Sigma(x-\bar{x})^2} \quad \text{...(2) ...(3)}$$

where $\bar{x} = (\Sigma x)/n$

and all the summations in these and other relations (to be given below) are performed from 1 to n (the no. of observations).

The 95% confidence Limit L_a for the slope 'a' may be obtained from the well-known Student's t-test (Wonnacott and Wonnacott, 1981). This limit is given by the relation

$$L_a = \pm \frac{st}{\Sigma\,(x-x)^{21/2}} \quad \text{... (4)}$$

where t_n is the student's t-static corresponding to 'n' (to be obtained from statistical tables) and 's' is the standard deviation of 'y' given by the relation

$$S = \left[\frac{\Sigma\,(y-y)^2}{(n-2)}\right]^{\frac{1}{2}} \quad \text{...(5)}$$

where $y = (\Sigma y) / n$. ... (6)

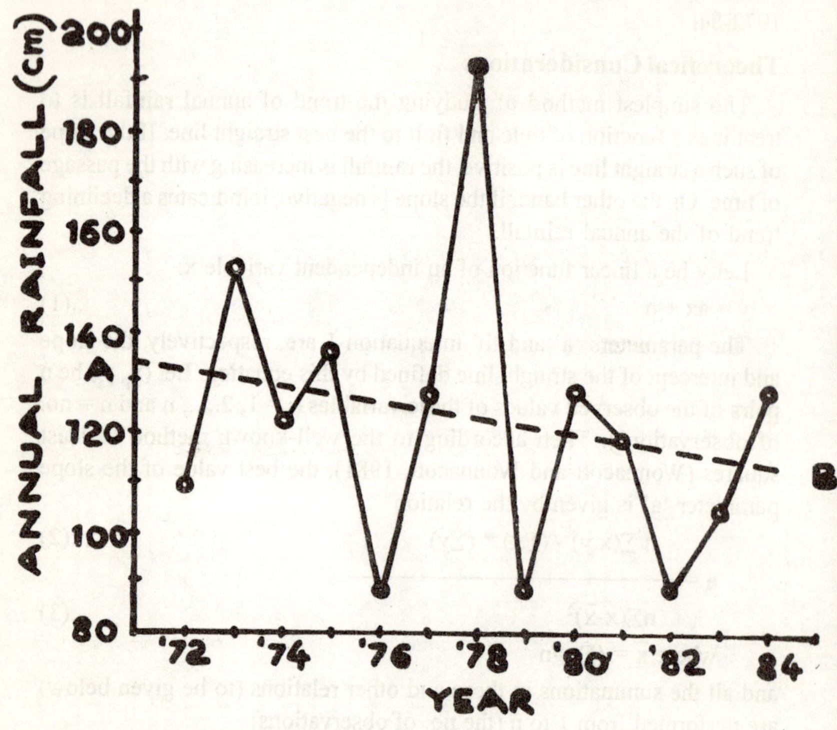

Fig. 19.1 : Average annual rainfall in Purulia District and its declining trend (shown by the best-fit line AB given by the relation $y = -1.608x + 133.7$)

An important parameter in connection with the annual rainfall, especially in the case of dry-land farming, is the relative interannual variability (IAV) of the rainfall. The obvious reason is that if the rainfall in a particular year deviates too much from its average value, the crops are affected adversely and the agricultural productivity suffers as a consequence.

Analysing the data on the rainfall and agricultural production in the Middle-Eastern countries, and correlating them, Wallen (1967, 1968) developed a precise formula for calculating the *relative* interannual variability (IAV) of the annual rainfall. According to Wallen,

$$\text{IAV (\%)} = \frac{100\left[\sum_{i=2}^{i=n} (R_{i-1}\ R_i)\right]}{(n-1)/\overline{R}} \qquad \ldots(7)$$

where R = annual rainfall, $\overline{R}$ = its average value, n = no. of years over which rainfall data have been collected and the sign indicates the numerical differences, disregarding its algebraic sign ≅.

Results and Discussion

The theoretical relations mentioned in the previous section have been applied to the annual rainfall data (Bhattacharjee and Chanda, 1988), presented in Table 19.1, for the district of Purulia. To perform the extensive numerical calculations required for this work, a FORTRAN Program was developed and run on a fast digital computer VAX-11/ 780 available at the Computer Centre of Regional Engineering College Rourkela.

The slope parameter 'a' was calculated using equation 1 to 3 by taking annual rainfall as the dependent variable 'y' and year as the independent variable 'x'. Equations 4 to 6 were used to calculate the 95% confidence limits L_a in each case. Our final results for 'a' and L_a for the ten rainguage stations in Purulia district are presented in Table 19.2. The last row of this table, shows the values of 'a' and L_a for the whole district, based on the annual average values of rainfall taken over ten rainguage stations of the district. Similarly, the relative inter-annual rainfall for each rainguage station and the whole district was calculated using equations 7, and the results are given in Table 19.3.

It is noted from Table 19.1 that the average annual rainfall in the district of Purulia for the 13-year period (1972-1984) is 122.4 cm. However, there is a wide fluctuation from year to year, ranging from

Table 19.2 : Best values of the slope parameter a and its 95% confidence limits ± L_a for the straight line y = ax + b (y = annual rainfall, x = year)

Station	$a \pm L_a$
Raghunathpur	-.764 ± 10.57
Kashipur	1.287 ± 5.313
Hura	-2.083 ± 15.36
Man Bazar	-0.850 ± 25.60
Purulia Town	4.717 ± 10.84
Bara Bazar	-11.28 ± 32.84
Hatwara	-2.026 ± 4.722
Bagmundi	0.964 ± 14.81
Jhaldah	-5.126 ± 5.322
Para	-3.405 ± 13.05
Purulia District	-1.6.08 ± 4.992

Table 19.3 : Relative interannual variability (IAV) for the annual rainfall in Purulia District

Station	*IAV (%)*
Raghunathpur	39.6
Kashipur	36.7
Hura	44.1
Man Bazar	24.7
Purulia Town	28.8
Bara Bazar	35.9
Hatwara	20.9
Bagmundi	29.1
Jhaldah	30.2
Para	39.8
Purulia District	31.9

62.5 cm (at Jhaldah in 1983) to 238.1 (at Hura in 1978). Even the average annual rainfall for the whole district ranges from 88.5 cm (in 1976) to 194.4 cm (in 1978). These wide fluctuations are also reflected in the relative interannual variability (IAV) of the annual rainfall. We find from

Table 19.3 that the relative IAV for the whole district is 31.9%, while it ranges from 24.7% (for Man Bazar) to 44.1% (for Hura). These wide fluctuations in the annual rainfall made Purulia a drought-prone district.

A close scrutiny of Table 19.2 shows that the values of 'a' are positive only at three rainguage stations, viz., Kashipur, Purulia town and Bagmundi, and negative at the remaining seven stations, it is hardly surprising, therefore, that the value of 'a' for the whole district is negative (Fig. 19.1, Table 19.2). Moreover, when we take into account the magnitudes of the slope parameter 'a', we find from Table 19.2 that the positive values of 'a' are usually smaller than the negative values. Thus the present analysis indicates that the annual rainfall in Purulia district has been slowly declining at seven of its ten rainguage stations during the 13 year period (1972-1984), probably as a result of large-scale deforestation and other ecological and environmental disturbances. Moreover, this rate of decline for the district is about 1.6 cm. per year (Table 19.2).

However, when we examine the numerical magnitudes of the 95% confidence limits L_a, we note from Table 19.2 that the numerical values of L_a are always larger than the corresponding values of 'a', without a single exception. This clearly indicates that the conclusion regarding the declining trend of annual rainfall in Purulia district is, at best, a tentative one, since the uncertainties in the values of 'a' are quite large. In order to reduce this uncertainty (i.e., the magnitudes of L_a), we find from equations 4 to 6 that the rainfall data should be collected and analysed for a much longer period (i.e., the value of n should be increased). Since Purulia is already a drought-prone district, as mentioned earlier, and any decline in its annual rainfall will have serious implications for agriculture, there is an urgent need to closely monitor the rainfall pattern and any changes in it in Purulia. The present investigation is a small step in this direction.

Acknowledgement

The authors are deeply grateful to Prof. S. Misra, Principal, Regional Engineering College, Rourkela, for research facilities, constant encouragement and full co-operation.

References

Bhattacharjee, B.K. and Chanda, B.N., (1988) Rainfall-ground water recharge relationship in Purulia, West Bengal, Proc. Nat. Sem. on Water Resources Development and Management, Rourkela, Feb. 10-13 (1988), pp. 258-74.

Patel, M.K. and Tiwari, T.N., (1991) Trends in rainfall at Rourkela (India) during pre-monsoon and post-monsoon periods. In Das, R.K. (Ed.), Technical Annual, V. 32, Institution of Engineers (India), Orissa State Centre, Bhubaneswar, pp. 74-78.

Tiwari, T.N., (1989A) Variability of the annual rainfall at Rourkela (India). Indian J. Env. & Agri. V. 4, pp. 93-101.

Tiwari, T.N., and Patel, M.K., (1989) Some trends in the annual rainfall at Rourkela (India). Indian J. Env. & Agri. V. 4, pp. 155-60.

Tiwari, T.N., Patel, M.K. and Pradhan, A.K., (1990) Annual rainfall in Phulbani Town, Orissa (India) : Some trends. Indian J. Env. & Agri. V. 5, pp. 21-26.

Wallen, C.C., (1967) Acidity definitions and their applications, Geografiska Annal. V. 49, pp. 367-83.

Wallen, C.C., (1968) Agroclimatological studies in the Levant. Proc. Reading Symposium, Paris, UNESCO, pp. 225-38.

Wonnacott, T.H. and Wonnacott, R.J., (1981) Regression : A Second Course is Statistics, John Wiley & Sons, New York.

20

Utility of Macroinvertebrates in biological water quality in some Central Indian Rivers

K.S. RAO, S. SHRIVASTAVA,
U. CHOUBEY
AND
N.K. DAD

Introduction

The concept of biological water quality monitoring by Saprobian system is still at an initial stage of study in India. Data published on temperate waters are yet to be tested in tropical conditions. While some bacteria, ciliates, algae, periphytonic, benthic macroinvertebrates and fish was suggested as indicator organisms (Hynes 1963, Cook 1976, Leibmann 1951, Chutter 1972, Patrick 1954, Fjerdingstad 1964, Sramek-Husek 1958, Gaufin & Trazwell 1956, Hellawell 1972, Jones 1978, Kolkwiz and Marsson 1908, 1909, Rao *et al.*, 1987, Rao and Shrivastava 1989, Sarkar and Krishnamoorthy 1977, 1981, 1985 US EPA 1973), some biologists have questioned the applicability of this system itself. Its application in tropical climate cannot be identical, in view of the acute seasonal impacts of these ecological realms. Unless further data is available from different rivers in tropical regions, the application of saprobian system to tropical water quality monitoring cannot be generalised. The present paper reports results on these lines from Chambal, Kshipra, Khan and Narmada Rivers, and discussed the possibility of applying saprobian system under Indian ecological conditions.

Materials and Methods

Observations on water quality monitoring in four rivers Chambal, Kshipra, Khan and Narmada at seven, four, five and nine selected study sites respectively were incorporated in this study (Table 20.1). This was done for a period of one and half years at different periods and the methods used for the analysis of various physico-chemical parameters are as per the procedures given in APHA (1971). Benthic macroinvertebrates and their quantitative computations were done for each river and comparative graphs drawn. Surber sampler and Ekman grab were used for organismal

Table 20.1 : Details showing the study sites of Chambal, Khan, Kshipra and Narmada

S. No.	*Name of Study site*	*Location*	*Description of locality*
	Chambal River		
1.	Chambal River near the Bridge at Nagda.	Chambal river enters Nagda urban area of this locality.	The water quality is comparatively good
2.	Area in Nagda just before joining of the effluents of rayon factory.	This locality is near the above locality.	Good Water Quality.
3.	Confluence area of effluent with Chambal River.	Pollution load in the form øf effluents join river Chambal here.	Represents extremely polluted locality with extending anoxic zones.
4.	Area after 20 KM of effluent union.	This study site re-presents rapid flow area of the river.	Represents flowing water where the self purification process is progressing.
5.	Ghatabillod area of river Chambal	This locality is near the city Ghatabillod underneath the Road bridge.	Represents study site at Ghatabillod city.
6.	20 KM from Ghatabillod.	This study site is an-other rapid flow locale.	Further river stretch showing a continuous improvement in water quality.
7.	40 KM from Ghatabillod. Area	This study is present near Khalghat bridge area.	Still further river stretch area showing improvement in Water quality.
2.	Khan River		
	1. Ravishankar Bridge	Locality at Indore Urban area.	Where quality is extremely poor due to heavy inputs of effluentes from several textile factories.
	2. Sanwer Bridge	Another down stream - locality 30 K.M. away from Indore.	Area showing partial recovery in river quality due to self purification.
	3. Panthpiplai	A Study site in rural area where the river gets dried in Summer.	Water quality poor in Summer but better in rainy and winter season.
	4. Triveni	A Study site where Khan river joins with Kshipra.	Water remains throughout the year and represents the deepest part of the river. Pollution load is there but becomes lesser due to self purification.

3.	Kshipra River		
	1. Triveni	A study site where reservoir is constructed.	Water quality is comparatively good.
	2. Gowghat	This locality is situated at a distance of 4 KM from Triveni.	Represents extremely polluted locality.
	3. Ramghat	Locality at Ujjain Urban area.	Water Quality is poor in all months.
	4. Kalbhairav	This study site is located at a distance of 8 KM from Ramghat	Water quality is good.
4.	Narmada River		
	1. Punasa I	This study site is situated at a distance of about 10 KM from Punasa Village.	Water quality is good.
	2. Chhotatawa	This study site is situated at a distance of about 20 KM from Beer Village of Khandwa District.	Water quality is comparatively good.
	3. Omkareshwar	This study site is situated at a distance of about 2 KM from famous Mandhata island.	Water quality is good.
	4. Kaveri River	This study site is situated at a distance of about 1KM from the confluence point of Kaveri with Narmada.	Water quality is good.
	5. Mandleshwar	This station is located along Mandleshwar city.	Water quality is comparatively good.
	6. Maheshwar	This study site is located along the Maheshwar city area.	Water quality is good.
	7. Barwani	This study site is located at a distance of about 6 KM from Barwani.	Water quality is good.
	8. Koteshwar	This study site is located at a distance of 5 KM from Nisarpur Village of Dhar district.	Water quality is comparatively good.
	9. Navagaon	This study site is located at a distance of 5 KM from Navagaon village of Bharouch dist.	Water quality is good.

quantitative sampling. The samples were later preserved in 10% formalin. These organisms were identified and quantitated as per standard techniques (APHA 1971, Wetzel 1975, Vollienweider 1973). Shannon weaver index and saprobity indices (Wilhm and Dorris 1966, 1968) were derived and water quality was assigned as per Saprobian system (Kolkwitz and Marsson 1909, Wilhm and Dorris 1966, Sladecek 1961, 1963). A new method of deriving ratios indicating macroinverterae densities of monsoon, winter and summer of each collection site for each river on a 12 monthly basis was presented.

Results and Discussion

(A) Physicochemical Characteristics Pattern

Available preliminary studies on the physicochemical and biological factors of the rivers, under investigation are : Chambal (Dad 1981, Rao *et al.*, 1991), River Khan (Rao and Diwan 1977, Dad 1979, Rao and Shrivastava 1989), River Kshipra (Rao and Rao 1986) and Narmada river (Chatterji and Rao 1991). The present work is initiated to study the macroinvertebrate density and water quality as per details given in Tables 20.2, 20.3, 20.4, 20.5 and 20.6.

Table 20.2 shows details of seasonal fluctuation on a monthly basis of the physico-chemical parameters of Chambal river. The acute fluctuations in all the 14 parameters given can be seen. In Chambal river acute deterioration in water quality in study site III, where the D.O. levels on Monsoon, Winter and Summer represented 5.9 mg/1, 2.0 mg/1 and zero respectively. Similarly, the B.O.D. values in these seasons are 14 mg/1, 47 mg/1 and 193 mg/1. Same trend was recorded in other parameters like C.O.D., Alkalinity, Chloride, Hardness, Calcium, Phosphate, Silicate and Nitrate nitrogen. Table 20.3 shows the same data with regard to river Khan, Table 20.4 with regard to river Kshipra and Table 20.5 with regard to river Narmada. The physico-chemical characteristics pattern of all these rivers show the same trend with acute changes in water quality in summer and to a lesser extent in winter and rainy season. This trend shows that the changes in river water quality in tropical climate depends on :

1. The number of points of pollution load.
2. The amount of water volume in the river giving the perennial or the seasonal status.
3. The intensity of seasonal impact resulting in a reduction in the water volume.

Table–20.2 : Showing details of seasonal fluctuation in Physico-chemical parameters of Chambal river at Seven study sites

PARAMETERS	MONSOON							WINTER							SUMMER						
	I	II	III	IV	V	VI	VII	I	II	III	IV	V	VI	VII	I	II	III	IV	V	VI	VII
Temp °C	28.85	29.2	29.75	28.92	28.3	28.3	28.2	24.25	24.1	28.6	26.5	25.3	24.9	24.4	29.12	30.45	32.87	32.45	32.45	31.05	30.0
pH	7.9	8.2	7.2	7.5	7.8	8.2	7.9	8.2	8.3	7.1	7.3	7.6	7.9	8.3	8.5	6.75	7.0	7.3	7.6	8.1	8.1
D. O. mg/l	7.5	9.7	5.9	7.07	7.4	7.5	7.3	9.0	8.9	2.0	3.4	4.8	7.5	8.1	8.45	7.7	ab.	ab.	1.4	3.1	6.1
B.O.D. mg/l	4.6	6.55	14.35	8.8	8.9	6.9	4.9	5.2	3.8	47.2	37.6	25.9	10.5	5.7	4.55	6.65	193.5	134.6	91.05	55.7	22.4
C.O.D. mg/l	9.9	12.3	30.05	20.1	16.9	15.4	13.85	12.35	8.2	104.8	91.4	63.9	23.1	15.32	10.25	13.85	382.1	265.6	195.0	116.9	16.25
Free CO_2 mg/l	5.12	4.12	10.72	9.5	7.1	5.3	5.85	-	-	12.15	7.6	5.2	-	-	-	-	28.1	19.85	14.5	6.4	-
Carbonate mg/l	-	-	-	-	-	-	-	4.5	8.0	-	-	-	4.6	6.4	8.0	9.4	-	-	-	-	4.4
Bicarbonate mg/l	109.4	102.4	109.0	115.1	117.1	118.0	105.0	149.15	124.5	129.2	129.0	128.3	147.8	136.2	210.95	162.4	149.2	150.5	154.7	178.4	182.3
Chloride mg/l	24.35	18.35	54.7	36.82	27.4	25.2	21.8	33.45	33.7	99.8	89.32	68.65	51.3	43.3	34.62	44.3	320.2	279.2	211.3	156.0	64.5
Hardness mg/l	103.4	96.85	119.3	115.1	126.9	90.7	91.2	123.8	112.7	122.6	112.1	116.6	113.3	113.1	141.7	128.8	244.3	229.0	193.2	154.2	137.4
Calcium mg/l	22.0	20.95	22.32	27.2	20.75	19.5	20.2	39.2	35.0	65.2	56.5	43.8	37.6	35.7	46.12	44.5	139.5	144.8	113.4	83.2	65.7
Phosphate mg/l	0.03	0.06	0.18	0.23	0.09	0.09	0.10	0.038	0.04	0.03	0.03	0.02	0.023	0.025	0.42	0.16	0.10	0.25			0.05
Silicate mg/l	-	-	-	-	-	-	-	-	-	-	-	-	-	-	-	-	-	-	-	-	-
Nitrate Nitrogen	0.22	0.20	0.22	0.17	0.33	0.18	0.06	0.07	0.08	0.08	0.07	0.07	0.06	0.06	0.05	0.01	0.03	0.05	0.05	0.05	0.05

Table 20.3 : Showing details of Seasonal fluctuation in Physico-chemical Parameters of Khan river at five study sites

PARAMETERS	MONSOON					WINTER					SUMMER				
	I	II	III	IV	V	I	II	III	IV	V	I	II	III	IV	V
Temp °C	27.3	25.5	23.5	24.7	23.0	27.5	26.5	21.8	22.0	21.3	35.3	33.4	31.5	32.5	31.5
pH	7.2	7.2	7.8	7.9	7.6	8.0	7.9	8.1	8.1	7.9	8.9	8.9	8.3	8.2	8.0
D. O mg/l	Nil	4.2	6.0	7.2	7.6	0.25	4.6	6.9	8.3	8.9	Nil	2.5	5.8	7.0	8.5
B.O.D. 20°C mg/l	252.0	106.5	48.5	34.2	12.0	212.6	129.4	42.6	23.5	9.9	32.5	228.5	78.3	56.4	12.8
C.O.D. mg/l	412.2	200.4	59.0	70.3	26.4	428.3	149.5	59.4	52.4	20.4	625.0	425.2	120.5	78.5	22.5
Carbonate mg/l	Nil	Nil	Nil	Nil	Nil	Nil	0.5	0.8	1.0	2.3	Nil	0.7	1.2	1.8	5.0
Bicarbonate mg/l	108.4	105.4	90.4	80.3	74.0	339.2	228.4	169.0	112.3	108.3	395.6	294.2	225.0	185.4	162.0
Chloride mg/l	45.8	29.5	32.0	27.2	26.0	102.0	84.6	63.4	49.3	32.5	164.0	124.5	110.5	98.6	78.3
Phosphate mg/l	0.9	0.6	0.2	0.1	0.06	0.96	0.8	0.15	-0.08	1.5	1.3	1.0	0.5	0.09	
Nitrate mg/i	0.3	0.19	0.2	0.15	0.1	0.2	0.1	0.3	0.2	0.10	0.3	0.35	0.5	0.5	0.15

Table 20.4 : Showing Seasonal fluctuation in Physico-chemical Parameters of Kshipra river at five selected study sites

PARAMETERS	MONSOON					WINTER					SUMMER				
	I	II	III	IV	V	I	II	III	IV	V	I	II	III	IV	V
Temp °C	25.9	25.7	25.8	25.0	25.7	20.1	20.2	21.2	21.3	21.3	24.7	25.0	25.4	25.8	25.9
pH	7.9	7.5	6.9	6.9	7.9	8.15	8.0	7.4	7.4	8.2	8.3	8.0	7.0	7.0	8.2
Dissolved Oxygen	8.2	8.0	7.9	7.3	8.0	8.9	8.9	8.2	7.6	9.7	8.9	8.0	7.4	7.0	8.5
B.O.D. 20°C mg/l	-	-	-	-	-	10.0	26.0	40.0	40.0	-	20.0	27.0	45.0	45.0	-
Carbonate mg/l	5.5	4.2	7.4	9.12	7.25	5.4	5.3	9.7	13.2	7.9	6.7	7.5	10.2	12.8	12.2
Bicarbonate mg/l	108.5	95.5	91.7	97.00	100.7	147.0	130.5	110.0	118.0	127.0	154.0	198.2	146.2	161.2	179.5
Chloride mg/l	148	17.6	40.6	40.1	19.7	15.2	14.8	38.5	38.6	21.4	20.6	19.5	46.6	47.0	27.2
Hardness mg/l	122.5	119.7	129.7	132.0	123.1	143.7	143.7	149.5	149.0	144.5	179.0	182.0	186.5	186.7	180.3
Calcium mg/l	26.2	14.7	27.0	26.8	24.3	34.0	33.7	34.4	34.5	26.7	40.0	40.8	46.4	44.7	35.7

Station
I : Triveni
II : Gowghat
III : Ramghat
IV : Kalbhairav
V : Kaliadeh

Table 20.5 : Seasonal fluctuation in Physico-chemical Parameters of Narmada river at Nine selected study sites

PARAMETERS	MONSOON									WINTER									SUMMER								
	I	II	III	IV	V	VI	VII	VIII	IX	I	II	III	IV	V	VI	VII	VIII	IX	I	II	III	IV	V	VI	VII	VIII	IX
Temp °C	23.7	26.7	25.5	26.0	25.7	24.8	25.2	25.2	25.1	23.1	20.5	29.6	23.6	23.9	23.0	19.8	24.4	23.0	26.7	29.8	27.2	27.6	26.7	26.7	24.0	25.9	24.1
Transparency cm.	48.0	44.7	46.5	40.5	40.5	32.0	41.0	44.5	43.2	183.5	146.5	132.0	107.7	124.5	93.5	136.7	139.5	141.5	193.2	147.5	127.0	117.0	113.0	116.2	101.0	139.6	98.0
pH	78	8.0	7.9	8.0	8.0	7.9	7.8	8.0	8.0	7.5	7.5	7.5	7.5	7.5	7.4	7.5	7.6	7.6	7.9	7.9	7.9	8.0	7.9	7.7	7.9	8.0	8.2
D. O mg/l	8.7	8.3	7.7	7.2	7.3	7.4	7.8	8.3	8.7	12.9	13.7	11.9	11.3	10.9	7.4	12.1	11.7	13.0	8.1	7.7	8.5	7.5	8.2	7.7	8.8	8.9	8.6
B.O.D. mg/l	10.3	12.3	11.1	14.1	11.6	11.6	13.2	13.4	13.5	5.5	6.1	5.6	6.3	5.7	5.9	5.5	6.0	5.8	8.1	8.7	8.05	9.4	8.3	8.7	9.2	9.4	9.3
C.O.D. mg/l	20.0	23.1	20.3	23.6	19.0	22.2	21.4	21.8	22.2	16.8	18.40	16.2	18.6	16.9	17.8	17.1	17.9	17.2	16.8	18.8	17.2	19.5	16.9	17.4	17.8	17.9	17.3
Free CO_2 mg/l	-	-	-	-	-	-	-	-	-	-	-	-	-	-	-	-	-	-	-	-	-	-	-	-	-	-	-
Carbonate mg/l	25.0	25.5	23.0	25.5	22.5	22.5	29.3	25.0	22.5	23.0	28.5	21.5	27.0	18.0	13.5	22.0	19.0	17.5	21.5	24.5	21.0	23.0	17.0	21.5	26.5	26.5	23.5
Bicarbonate mg/l	173.5	200.5	180.5	177.0	176.0	170.0	181.0	173.0	178.0	164.0	212.0	179.5	171.5	161.5	172.0	180.0	173.0	185.0	183.0	175.0	183.0	175.0	173.0	171.0	179.0	170.0	172.0
Chloride mg/l	17.6	20.4	20.4	20.9	19.1	19.1	18.9	20.6	19.1	12.1	14.4	14.1	16.1	14.4	14.4	14.9	14.4	14.6	15.1	16.4	15.4	16.4	15.4	14.6	15.1	14.0	16.1
Hardness mg/l	174.5	182.0	166.0	172.0	170.0	168.0	160.0	171.0	179.0	165.0	187.0	156.0	165.0	158.0	160.0	139.0	169.0	161.0	167.0	172.0	162.0	168.0	162.0	165.0	160.0	167.0	171.0
Calcium mg/l	27.8	28.7	20.8	18.8	27.2	26.4	26.4	28.3	27.8	28.1	22.4	25.1	25.9	24.4	24.7	21.8	22.6	24.9	24.6	24.9	23.8	21.2	22.7	23.8	24.0	23.0	26.4
Phosphate mg/l	0.16	0.16	0.18	0.20	0.14	0.18	0.21	0.18	0.17	0.01	0.04	0.06	0.06	0.03	0.07	0.05	0.06	0.06	0.08	0.11	0.11	0.15	0.09	0.13	0.15	0.13	0.14
Silicate mg/l	13.7	14.0	13.3	13.5	13.0	13.0	13.0	12.7	13.2	13.0	10.2	11.8	8.7	11.2	9.2	9.0	9.5	11.3	11.1	10.7	11.7	11.5	12.1	12.6	12.4	13.0	13.1
Nitrate Nitrogen	6.0	6.5	7.0	6.9	6.35	6.5	7.0	6.7	6.3	7.0	6.7	4.6	5.3	3.7	3.9	4.0	5.7	5.0	4.5	4.4	5.4	5.0	4.7	5.7	7.4	5.1	5.3

Station
I : Punasa
II : Chhotatawa
III : Omkareshwar
IV : Kaveri
V : Mandleshwar
VI : Barwani
VII : Koteshwar
VIII : Navgaon

Table 20.6 : Showing macrozoobenthic Population density and zonation at the study sites of Chambal, Khan, Kshipra and Narmada rivers.

S. No.	*Name of River*	*Saprobic designation*	*Community*	*Remarks*
1.	Chambal River			
	Station I & II	Oligosaprobic	Ephemeropetra, Trichoptera and Mollusca	Diptera, Hirudinea Oligochaete community increase in summer.
	Station III	Polysaprobic	Tubifex, Chironomus	Oligochaete increased 10 times in summer than winter.
	Station IV & V	Mososaprobic	Tubifex, Chironomus	- Do -
		Station VI & VII	Oligosaprobic	Ephemeropetera, Trichoptera No showing fluctuations in season.
2.	Khan River			
	1. Ravishankar Bridge	Polysaprobic	Tubifex & Limnodrilus	Highest density occurred in summer.
	2 & 3. Bawan Pipliya & Sanwer Rd.	a-Mesosaprobic	Tubifex, Chironomus & Mollusca Bridge.	Oligochaete & Diptera % increased in winter.
	4. Panthpipliya	b-Mesosaprobic	Chironomus Mollusca & Ephemeroptera	Oligochaete & Diptera in winter.
	5. Triveni	Oligosaprobic	Trichoptera, Ephemeroptera and Diptera	In winter Trichopteran % increased when compared to Ephemeroptera.

Table 6 (Contd..)

S. No	*Name of River*	*Saprobic designation*	*Community*	*Remarks*
3.	Kshipra River			
	1. Triveni	Oligosaprobic	Ephemeropetra, Trichoptera and Diptera.	Normal pattern.
	2. Gowghat	b-Mesosaprobic	Mollusca Chironomus	Species diversity increased in summer.
	3. Ramghat	Polysaprobic	Chironomus, Tubifex, and Lymnaea.	Chironomous population increased and Mollusca reduced in summer.
	4. Kalbhairav	Polysaprobic	Chironomus, Tubifex and Lymnaea.	Lymnaea eliminated in Summer.
4.	Narmada River			
	1. Punasa I & II	Oligosaprobic	Ephemeroptera, Diptera and Mollusca.	In summer months it becomes a-Mesosaprobic.
	2. Omkareshwar	Oligosaprobic	Ephemeroptera, Plecoptera and Diptera	a-Mesosaprobic condition on January.
	3. Mandleshwar	Oligosaprobic	Ephemeroptera, Diptera	a-Mesosaprobic in summer. and Annelida.
	4. Maheshwar	b-Mesosaprobic	Ephemeroptera, Mollusca Diptera and Oligochaeta	Oligosaprobic condition in winter.
	5. Rajghat	Oligosaprobic	Ephemeroptera, Mollusca and Diptera.	a-Mesosaprobic condition in summer.
	6. Koteshwar	b-Mesosaprobic	Ephemeroptera, Diptera	Oligosaprobic condition in and Mollusca winter.

4. The water evaporation and the porosity levels. (Water retention capacity) of the river bed.
5. River substratum, terrain, water velocity, gradient and rates of aquatic macrophyte infestation and the nature of the river basin.

Looking at the above points the present four rivers under study may be further categorised as :

(A) Water quality of rivers receiving high pollution load :

Under this category river Chambal and river Khan may be included. As per Tables 20.2 and 20.3 the Physico-chemical characteristics in these rivers show high water quality deterioration at the points of pollution inputs, like station III Chambal river at Nagda receiving the Grasim Pollution load and station I on river Khan receiving effluents from textiles factories at Indore. The selfpurification process in these rivers need further intensive studies but depending on the present base the selfpurification level in Chambal is comparatively better than river Khan, where it remains polluted throughout its course. This may be related to the amount and points of pollution inputs in these respective rivers. The pollution inputs on river Chambal are comparatively lesser than those of river Khan. The physico-chemical data depicts the same pattern. The greater reduction in water quality in river Khan could be due to multiple reasons like more pollution inputs, less water volume, less water velocity, greater evaporation. Water in this river is retained in areas having rocky basin. These aspects need further study.

(B) Water quality of river receiving medium pollution load :

Under this category Kshipra river may be included in which the pollutants include effluent load from smaller industries and high urban organic pollution load. Table 20.4 gives the seasonal physicochemical characteristics of the river. While the present study restricts its purview to saprobian system no detailed investigations on the inorganic contaminants were reported. The general physico-chemical parameters at the five study sites due to sewage, domestic wastes and intense human bathing denote highest organic pollution inputs at Ramghat and Kalbhairav when compared to other study sites. The pattern of the physico-chemical characteristics in parameters like pH, Dissolved Oxygen, B.O.D., Alkalinity, Chloride, Hardness and Calcium denote the same pattern. These also exhibit higher water quality deterioration in summer when compared to other stations.

(C) Water quality of rivers receiving less Pollution load :

Under this category river Narmada which receives the least pollution load may be included. Table 20.5 shows the extensive Physico-chemical data at IX study sites in the western Zone of this river. The physico-chemical characteristics pattern in this river shows a uniform trend in water quality at all the study sites. One of the main reason for the better water quality of this river, seems to be lesser inorganic pollution load, quicker, shedding or organic pollutants in the river especially in Omkareshwar and Maheshwar stretch and the perennial nature of the river showing more water volume.

(B) Biological Characteristic Pattern

Studies on qualitative and quantitative macrozoobenthic organisms form the important criteria for evaluation, leading to water quality designation according to saprobian system. The preliminary data available on these lines on river Chambal (Rao and Shrivastava 1989, Rao *et al.*, 1991), river Kshipra (Rao and Rao 1986) and River Narmada (Chatterji and Rao 1991), is insufficient and fragmentary. The present study giving the macrozoobenthic densities at the study sites have made community evaluations and designations of water quality on the basis of the saprobian system.

In Chambal river study sites I and II were given the oligotrophic status showing community zones of Ephemeroptera, Trichoptera and Mollusca. Besides, Diptera, Hirudinea and Oligochaeta which increase in their densities in summer were also pointed out.

Study site III (Table 20.6, Fig. 20.2B) was designated as showing the polysaprobic water quality with the dominance of *Tubifex* and *Chironomus* community. Station IV and V showed the mesosaprobic water quality levels and St. VI and VII showed the oligosaprobic status was pointed. In river Khan, study site I at Ravishankar bridge was given the polysaprobic designation with dominant communities of *Tubifex* and *Limnodrilus* showing highest densities in summer. Study site II and III at the Baman Pipliya and Sanwer Localities showing dominant *Tubifex, Chironomus* and Mollusca communities were given the mesosaprobic status. Study site IV Panthpipliya showing *Chironmous,* and Mollusca and Ephemeroptera communities was also given a water quality status of mesosaprobic level. Study site V at Triveni was designated the Oligosaprobic status showing Trichoptera, Ephemeroptera and Diptera communities. In Kshipra river, the Triveni locality was designated the oligosaprobic status, Gowghat mesosaprobic status and Ramghat and

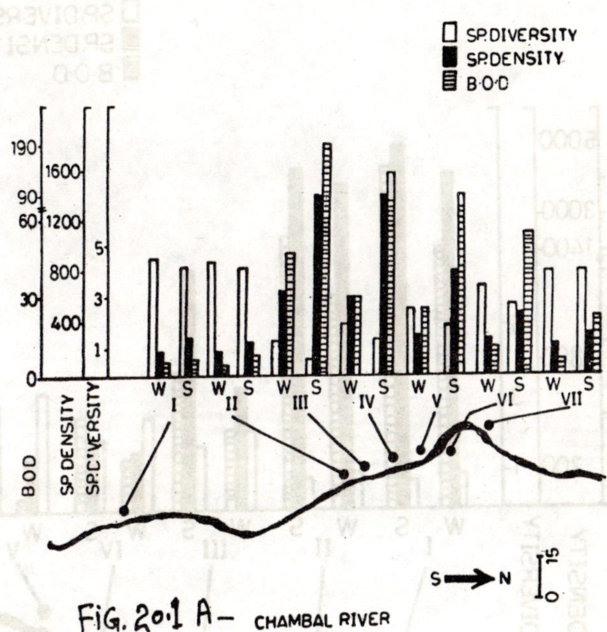

FIG. 20·1 A– CHAMBAL RIVER

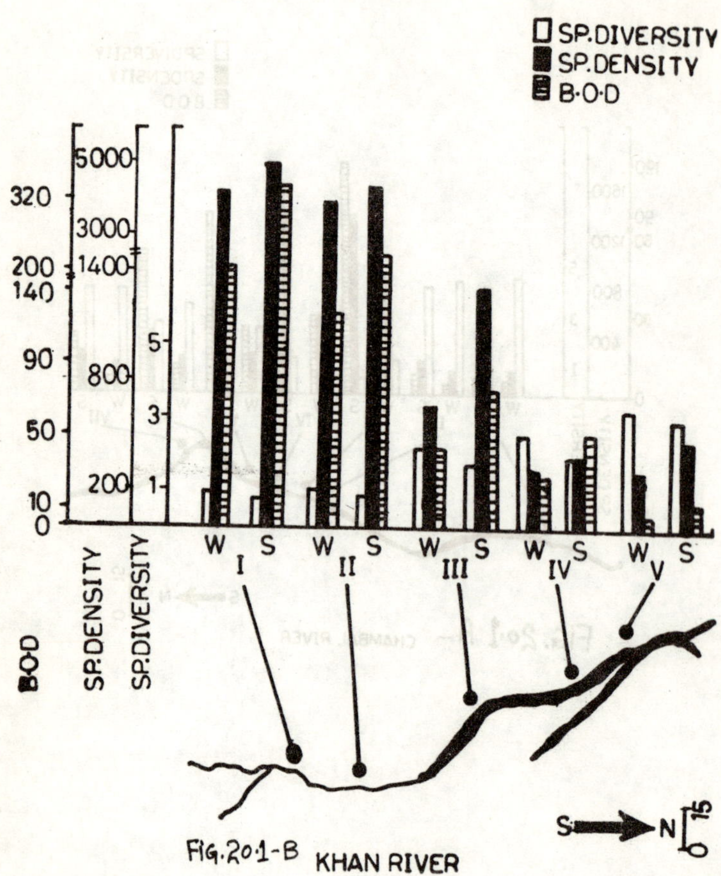

FIG.20.1-B KHAN RIVER

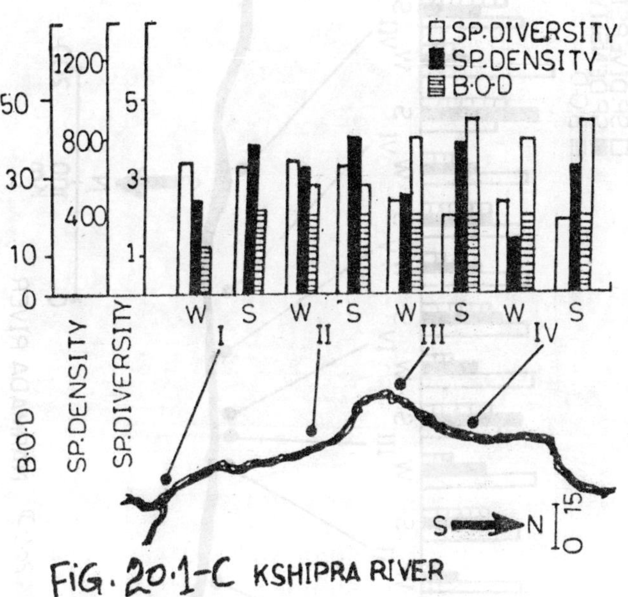

FiG. 20·1–C KSHIPRA RIVER

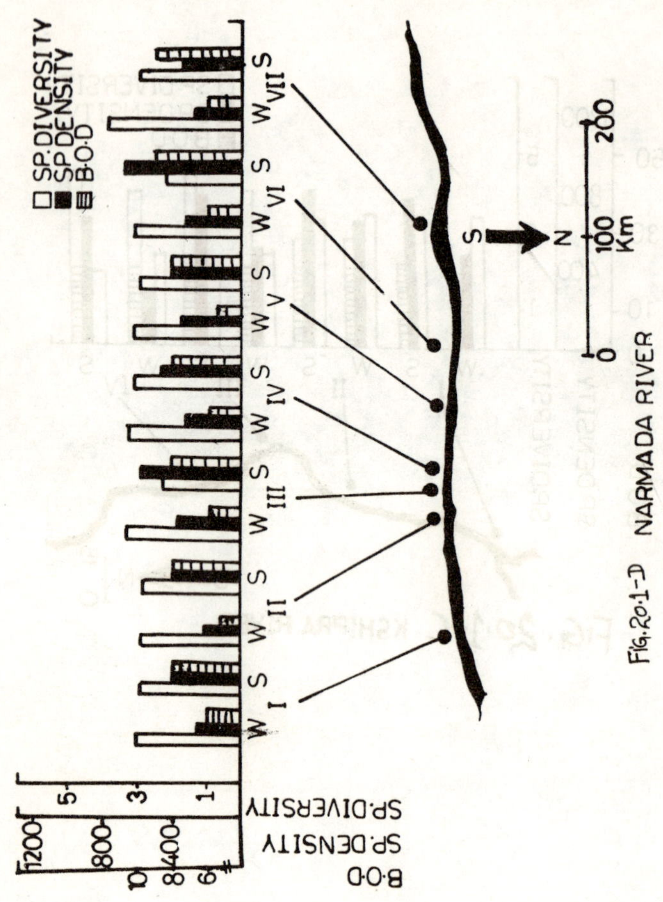
SP. DIVERSITY
SP. DENSITY
B.O.D
1200
800
400
10
8
6
5
3
1
W S W S W S W S W S W S W S
I II III IV V VI VII
S
N
0 100 200
Km
NARMADA RIVER

Fig. 20.1-D

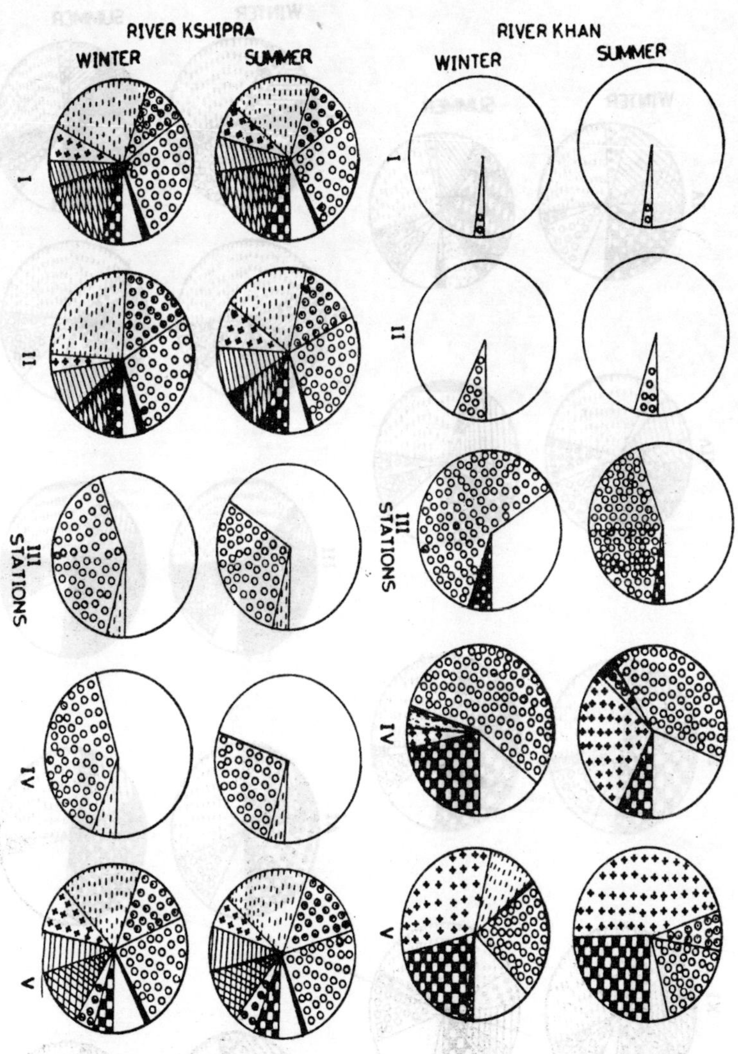
RIVER KSHIPRA
WINTER
SUMMER
RIVER KHAN
WINTER
SUMMER
I
II
III
STATIONS
IV
V
I
II
III
STATIONS
IV
V

Fig. 20.2

RIVER NARMADA

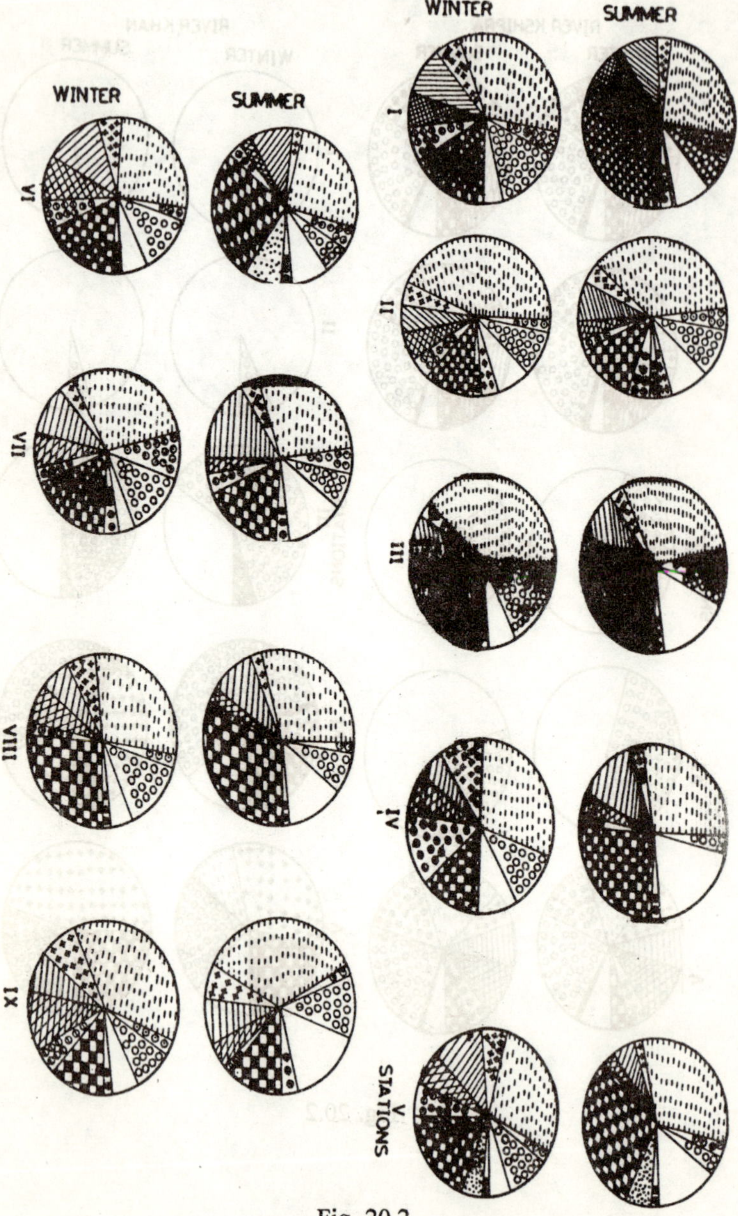

Fig. 20.2

RIVER CHAMBAL

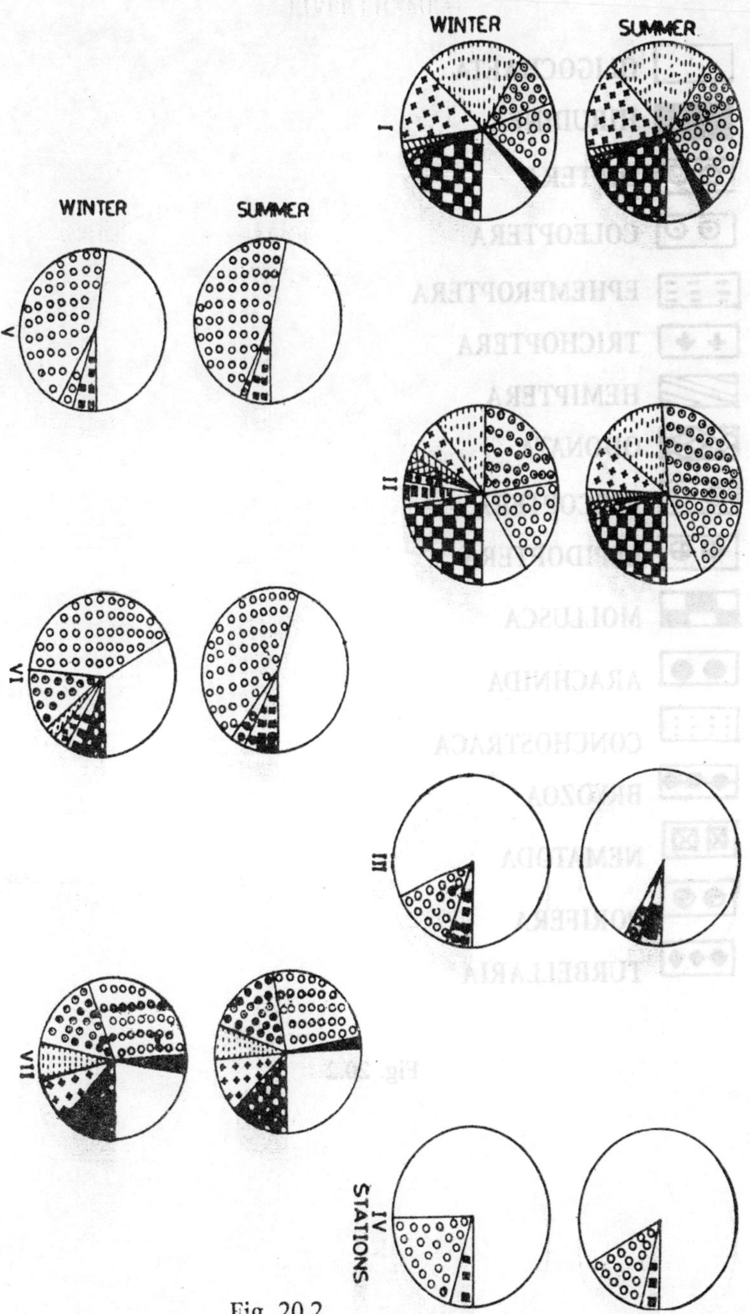

Fig. 20.2

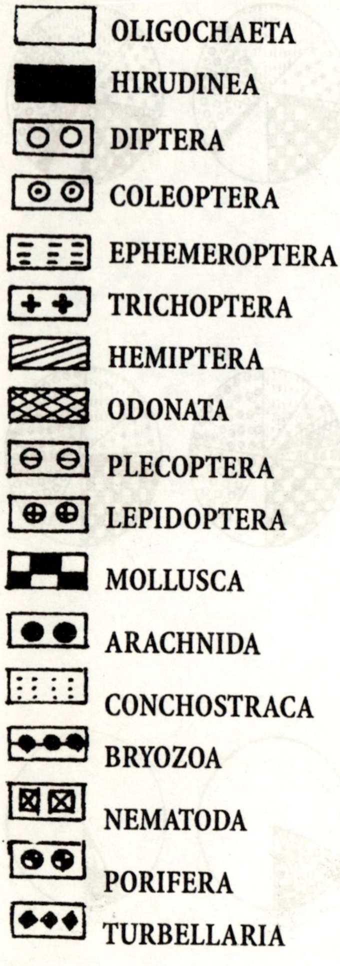

OLIGOCHAETA
HIRUDINEA
DIPTERA
COLEOPTERA
EPHEMEROPTERA
TRICHOPTERA
HEMIPTERA
ODONATA
PLECOPTERA
LEPIDOPTERA
MOLLUSCA
ARACHNIDA
CONCHOSTRACA
BRYOZOA
NEMATODA
PORIFERA
TURBELLARIA

Fig. 20.2

Kalbhairav designated the polysaprobic status. It was observed that the Molluscan density is comparatively higher at all the localities and the *Lymnaea* population was eliminated in summer months. In Narmada river all the localities except Maheshwar (B Mesosaprobic) and Koteshwar (B Mesosaprobic) were designated to be having the oligosaprobic status. The Ephemeroptera, Plecoptera and Mollusca community dominated the entire western stretch of Narmada river.

The above designated categories of water quality monitoring show the better quality of Narmada > Kshipra > Chambal > Khan. In Fig. 20.2 E derived saprobic index values for river Narmada, Chambal, Kshipra and Khan rivers were given. Fig. 20.1 A, B, C, D, show quantitative histograms representing the macrobenthic species diversity and density and the B.O.D. values, while Fig. 20.2 A. B. C. D., represent the relative dominance of various groups of macrozoobenthic organisms in rivers Narmada (A), Chambal (B), Khan (C) and Kshipra (D).

It can be seen that the self-purification levels are applicable to both organic and inorganic contaminants. While organic pollution can be easily monitored, the inorganic contaminants need elaborate analysis. Hence it is suggested that the rates of self-purification process may be used as indicators to designate water quality, because this applies to both organic and inorganic contaminants.

Special consideration was accorded in the present study to evaluate seasonal fluctuations in the water qualities in the four rivers under consideration. The seasonal impact factors computed in order of monsoon, winter and summer for different rivers, based on density diversity ratio range from 1:1.6:2 for river Chambal, 1:1.5:2 for river Khan, 1:3:5 for river Kshipra and 1:3:5 for Narmada river. It is interesting to note that the earlier studies by Rao *et al.*, (1987) derived 1:1.5:2 as the seasonal impact ratio on the basis of their study on fresh water Bryozoa at various lentic and lotic aquatic bodies of central India. The ratios pointed out in the present study for Kshipra, Khan, Chambal and Narmada rivers broadly agree with this. It would be possible to standardise these ratios of seasonal impact on a generalised basis.

Acknowledgements

The authors are grateful to Dr. G.N. Johri, Prof. and Head S.S. in Zoology, Vikram University, Ujjain (M.P.) for the facilities provided.

References

American Public Health Association, (1971) Standard Methods for the examination of Water Sewage and Industrial Wastes. New York.

Cook, S.E.K., (1972) Quest for an index of community structure sensitive to water pollution. Environ. Pollut., V. 11, pp. 264-68.

Chutter, F.M., (1972) An empirical biotic index of the quality of water in south African Streams and rivers. Limnol, Ocea, V. 17, pp. 139.

Chatterji, S.N. and Rao, K.S., (1991) Studies on the preimpoundment fisheries potential of Narmada river at Western Zone, J. Inland Fisheries CIDFRI, Barrackpore. (in press).

Dad, N.K. (1979) Studies on Khan River with Special reference to Pollution M. Phil Thesis, submitted, 1979.

Dad, N.K. (1981) Limnological studies on Chambal river with reference to pollution. Ph. D. Thesis, Vikram University, Ujjain, 227 pp.

Fjerdingstad, E., (1964) Pollution of streams estimated by benthal phytomicro-organisms 1-A Saprobic system based on Communities of organisms and ecological factors. Int. Rev. Ges Hydrobiol. V. 49, pp. 63-131.

Gaufin, A.R. and Tarzwell, C.M. (1956) : Aquatic macroinvertebrate communities as indicators or organic pollution in LyHe creek. Sewage and Industr. wastes 28, 906-24.

Hellawell, J.M., (1972) Biological surveillance of rivers. Water Research Centre, Stevenage, U.K.

Hynes, H.B.N., (1963) The Biology of Polluted waters. Liverpool, Univ, Press.

Jones, R.C., (!978) Algal locomass dynamics during colonization of artificial islands : Experimental results and a model. Hydrobiologia, Y. 59, pp. 165.

Kolkwitz, R. and Marson, M., (1908) Okologic der pflanztichen Saprobien, Ser. Dent. Bot. Gesell, V. 26, pp. 505-19.

Kolkwitz, R. and Marson, M., (1909) Okologie der tierische Saprobien. Inter. Rev. D. Ges. Hydrobiologie, U. Hydrographie, 2, 126-52.

Lackey, J.B. (1938) Protozoan plankton as indicators of pollution in a flowing stream. Publ. Health. Rep. Wash, V. 53, pp. 2037-58.

Leibmann, H., (1951) Handbuch der frischwasser and abwasser, Biologia, Bd. 1 ed. Munchen, Jena.

Patrick, R., (1954) Diatom as an indicator of river change. Proc. 9th Indus. Waste Conf. Pundus, Univ. Engne ser., V. 97 pp. 325-30.

Rao, K.S., and Diwan, A.P., (1977) Studies on Pollution, in Khan Nadi, Indore. Factors influencing the Ectoproct *Plumatella repens*

(Phylactolaemala: Plumatellidae) Proc. Sem. Eur. Eng. Div. Inst. Engineers S. III. 18. 1-18.4.

Rao, P.S. and Rao, K.S. (1986) Pollution due to pilgrim bathing during Kumbh Mela 1980, in Kshipra river. J. Hydiobio'. V. 2-4. pp. 47-55.

Rao, K.S. Diwan. A.P. Srivastava, P. Swarup Abha and Dhakad, N.K., (1987) Studies on Water quality monitoring with fresh water Ectoprocta as indicator organisms in Perspectives, in Hydriobiology Eds/ K.S. Rao and S. Srivastava Sec. III (21) pp. 99-117.

Rao, K.S. and Shrivastava, S., (1989) Studies on biological monitoring of Water quality in Chambal and Khan rivers of central India Geobios, V. 16, pp. 78-82.

Rao, K.S., Shrivastava, S., Choubey, U and Rao, B.S., (1991) Observations on biological monitoring concepts of some riverine ecosystem of central India, Proc. Nat. Sym. on New Horizons in Fresh Water Aqu. Jan. pp. 23-25.

Sarkar, R. and Krishnamoorthi, K.P. (1977) Biological Method for monitoring water pollution levels studies at Nagpur, Indian J. Environ, Hlh, V. 19, pp. 132.

Sakar, R. and Krishnamoorthi, K.P., (1981) Ecology of free swimming ciliates from stabilization ponds, Nagpur, India VI International Congress.

Sarkar, R. and K.P. Krishnamoorthi, (1985) The Current concepts of biological monitoring of inland waters IAWPC, Tech. Annual XII pp. 58-70.

Sladecek, V., (1963) Zur biologischen Gliederung der hoheren Saprobitatsstien. Arch. Hydrobiol, V. 58 (1), pp. 103-21.

Sladecek, V., (1963) The future of the Saprobity system in "Symposium on the evaluation", of the saprobity system. Mar. 25-27, 1963, pp. 518-27.

Sramek-Husek, R., (1958) Die rolle der cillatenanlyse bei der biologischen Kontrolle von flussverunnein igungen. verb. Int. ver. Limno., V. 13, pp. 639-45.

United States Environmental protection Agency, (1973) Water quality criteria 1972. Ecological Research series, EPA. R3 73.003 March 1973. Washington D.C. pp. 594.

Vollienweider, A., (1973) A manual on methods for measuring primary production in Aquatic environments, IBP Hand book V. 12 Blackwell Oxford.

Wetzel, R.G., (1975) "Limnology" W.B. Saunders Co. Philadelphia, U.S.A. 743. pp.

Wilhm, J.L. and Dorris, T.C., (1966) Species diversity of benthic macroinvertebrates in a stream receiving domestic and oil refinery effluent Am. Midl. Nat. V. 76, pp. 427-49.

Wilhm, J.L. and Dorris, (1968) Biological parameters for water quality criteria. Bioscience, V. 18, pp. 473.

21

Utilization Aspects of some Aquatic Plants

A. DEWANJI

Introduction

Water has always been an important prerequisite for development. In developing countries like India, water has great economic importance for its use in irrigation, hydropower, navigation, industries, fisheries as well as for domestic purposes. In spite of its importance, there has been little thought of caring for the bountiful supplies of water which seems endlessly renewable. With the advent of civilization and industrial development, large quantities of inorganic nutrients present in effluents from villages, farms and industries together with the runoff of fertilizers added to the field, enriched the water of canals and reservoirs. Because of the continued enrichment of waterbodies in limiting nutrients, particularly nitrogen and phosphorus, aquatic plant growth has been rapid in fresh waters (Aboaba, 1973). The accelerating growth of the aquatic weed flora, over the years, has now reached alarming proportions.

Harmful Aspects

The damaging aspects of aquatic weeds include blocking of canals and pumps in irrigation projects, interfering with hydroelectricity production, hindering boat traffic, increasing loss of water by evaporanspiration, aggravating the incidence of water borne diseases, interfering with fishing and fish culture, reducing recreational values of water ways, reducing yields in paddy crops, clogging rivers and canals so that drainage becomes impossible and floods result (Varshney and Singh, 1976). This is a global problem, but in the warm waters of the tropics and subtropics conditions for excessive development of these water plants are particularly favourable as growth is not even periodically retarded during a winter season.

As the serious negative implications of the presence of aquatic weeds have become widely recognized, various eradication measures are being employed to control excessive weed growths. In recent times, the need for aquatic plant utilization as opposed to their eradication with herbicides is receiving increased attention (NAS, 1984).

Utilization

The state of West Bengal in India has vast areas of waterbodies comprising of rivers, lakes, canals, ditches, ponds, tanks, swamps and marshes. Being a monsoon region, these water bodies favour the growth of all kinds of aquatic vegetation and have rich aquatic flora (Majumdar, 1965). The proper functioning of these waterbodies is often hampered by an excessive growth of aquatic weeds. Utilization of these unwanted plants, as an additional source of food and feed—a pressing need of our country, would also help to solve the weed eradication problem.

This paper deals with the potentially of using the local aquatic flora as a food and feed resource based on research conducted in our own laboratory. Work was conducted on lush, green samples of thirty aquatic weeds (Table 21.1) which were selected according to their availability and collected from sites in and around Calcutta.

Use as Animal Feed

Chemical composition

The nutritive value of any feed or feed mixture is its capacity for supplying organic and inorganic nutrients which the animals can actually use for maintenance and production. In a study on the chemical composition of the thirty weeds (Banerjee and Matai, 1990), twelve aquatic plants were found to possess a satisfactory composition when compared to commonly used roughages (Table 21.2).

From the Table, it is evident that aquatic plants had larger amounts of crude protein than the roughages. Protein happens to be an essential component of any diet and different classes of animals have varying nutritional demands for protein which are particularly high during late pregnancy and lactation. Since roughages such as straws which form the staple feed for the most of the Indian livestock contain less than 4 to 5% protein, an additional source of protein is generally needed with straw dites (Aitchinson, 1988). Since all the aquatic weeds contain more than 14% crude protein, these could provide an important source of supplementary protein in animal feeds. Dolberg *et al.*, (1981) reported an increase in the feed intake, nitrogen retention and weight gain of cattle when water hyacinth, azolla and duckweeds were used in combination with feeds of poorer quality such as rich straw. High ash values in some plants could be a limitation but if the plants were used to supplement feeds and not comprise the entire diet then the high mineral content of these plants would not pose a significant problem.

Table 21.1 : Aquatic plants with scientific, common names and month of collection that were collected near Calcutta for utilization studies

Scientific Name	*Common Name*	*Month of collection*
Allamania nodiflora R. Br.	–	July
Altepanthera philoxeroides	Alligator weed	August
Azolla pinnata R. Br.	Water velvet	September
Ceratophyllum demersum L.	Hornwort	October
Commelina benghalensis L.	Day flower	August
Eichhornia crassipes (Mart.) Solms.	Water hyacinth	October
Enhydra fluctuans Lour.	Watercress	September
Gentiana lutea	–	September
Hydrilla verticillata (L.) Royle	Hydrilla	August
Hydroctyle asiastica L.	Penny wort	September
Hygrophila spinosa	–	August
Impmoea reptans Poir	Water spinach	September
Jussiaea repens L.	–	March
Lemina minor L.	Duckweed	August
Limnathemum cristatum (Roxb.) Griseb.	Snowflake	July
Limnophila herophylla	–	September
Lippia geminata	–	August
Marsilea quadrifolia L. Willd	–	March
Mikania scandens Wild[1]	Climbing hempweed[1]	August
Nelumbo nucifera Gaertn.	Lotus	July
Nymphae nouchali Burm. f.	Red waterlily	July
Nymphaea stellata Willcd	Blue waterlily	July
Ottelia alismoides (L.) Pers.	–	September
Pistia stratiotes L.	Water lettuce	August
Phragmites karka (Retz.) Trin. ex. Stued.	Giant reed	September
Polygonum barbatum L.	Knot weed	August
Salvinia auriculata Aubl.	Water fern	July
Trapa bispinosa Roxb.	Water chestnut	October
Typha elephantina Roxb.	Cattails	October
Vallisneria spiralis L.	Tapegrass	March

[1] Milkinia scandens is basically a terrestrial weed but the sample used here grew profusely in damp ditches and on margins of ponds.

Table 21.2 : Proximate composition (% dry weight) of aquatic plants[a] and some commonly used roughages[b]

Plants	*Crude protein %*	*Crude fibre %*	*Crude fat %*	*Ash %*	*Nitrogen free extract %*
Aquatic plants[a]					
Allamania nodiflora	22.6	22.3	4.9	14.7	35.5
Azolla pinnata	21.9	16.6	3.8	15.4	42.3
Enhydra fluctuans	19.7	22.9	4.3	16.8	36.3
Eichhornia crassipes	14.5	20.8	4.3	13.2	47.2
Ipomoea reptans	20.6	27.0	4.6	16.5	31.3
Hydrilla verticillata	16.6	15.1	3.7	20.4	44.2
Lemna minor	20.4	15.7	3.8	17.2	42.9
Limnanthemum cristatum	24.4	22.4	2.2	14.5	36.5
Limnophila heterophylla	15.4	23.3	4.3	12.6	44.4
Lippia geminata	19.1	24.6	3.8	18.3	34.2
Marsilea quadrifolia	16.8	23.4	4.6	19.5	35.7
Pistia stratriotes	20.5	19.1	3.8	17.0	39.6
Roughages[b]					
Cowpea (green roughage)	18.2	25.3	2.6	14.2	39.6
Berseem (green roughage)	17.4	25.9	1.9	14.2	40.7
Cowpea hay	15.3	34.8	1.1	13.3	35.4
Berseem hay	14.7	30.6	1.6	12.1	41.0
Wheat straw	3.2	34.9	1.5	11.6	43.8
Paddy straw	3.4	33.6	1.2	18.4	42.6

[a] Source : Banerjee and Matai, 1990

[b] Source : Banerjee, 1988

Factors Affecting Quality

If aquatic vegetation is to be used as feed material, some attention should also be given to factors affecting the quality of the plant material. Many plants contain a wide variety of natural chemical compounds often referred to as anti-nutritive substances or toxic factors which are capable of inducing toxic effects in animals consuming these plants (Bondi and

Alumot, 1987). Some of the toxic factors present in the twelve aquatic plants are given in Table 21.3.

Alkaloids present in certain plants, many of them having poisoning properties, comprise the largest single class of secondary plant substances. Their qualitative presence was tested in plant samples subject to differential extraction (Hultin and Torsell, 1965). Their complete absence was noted in both fractions of ten plants. Nitrates are known to accumulate in forage plants and in some weeds at levels high enough to be toxic to animals (Wright and Davison, 1964). A nitrate content of less than 1.5% has been considered to be safe for use as animal feed (Bondi and Alumot, 1987). All the twelve aquatic plants had values below 1.5%. The phenolic contents of the plants were studied in view of the adverse effect of these compounds on growth due to their interference with protein digestibility (Bondi and Alumot, 1987). Total polyphenols ranged from 1% in *Hydrilla verticillata* to 7.2% in *Lemna minor*.

Table 21.3 : Potential toxic compounds (% dry weight) found in some aquatic plants collected near Calcutta[a]

Plants	*Alkaloids Fraction* A	*Total Fraction* B	*Nitrates* %	*Polyphenols* %
Allamania nodiflora	-	-	1.2	2.9
Azolla pinnata	-	-	1.3	5.2
Enhydra fluctuans	+	-	1.4	5.0
Eichhornia crassipes	-	-	0.8	4.1
Ipomoea reptans	-	-	1.4	3.7
Hydrilla verticillata	-	-	1.2	1.0
Lemna minor	-	-	0.8	7.2
Limnanthemum cristatum	-	-	1.0	3.2
Limnophila heterophylla	-	-	0.8	3.6
Lippia geminata	-	-	1.3	5.2
Marrsilea quadrifolia	+	-	1.3	3.3
Pistia stratriotes	-	-	1.3	2.2

[1] Fraction A was the part extracted with chloroform from an ammonical solution and Fraction B was the part extracted with chloroform : ethanol (3 : 2) after addition of sodium sulphate. (-) absence ; (+) presence.

[a] Source : Banerjee and Matai, 1990.

Thus, these twelve aquatic weeds were found to contain sufficient quantities or nutrients and were safe to be considered as potential livestock feed. Further studies on other aspects of feed quality such as palatability and digestibility experiments should be initiated on these plants.

Use as Food

Most aquatic plants cannot be consumed directly because of the presence of secondary constituents in them (McClure, 1970), but the extraction of edible protein from their leaves could prove promising. The mechanical separation of protein from leaves is based on the principle of rupturing the cell wall, pressing out the cellular fluid containing the protein bearing plastids and finally coagulating out the proteins. The possibility of using such extracted leaf protein, free from fibrous mass and toxicants, as dietary supplements to improve quality and raise protein levels of diet has been suggested (Pirie, 1987).

Trials using thirty, locally harvested aquatic weeds were conducted to study the feasibility of using these plants as a possible source of leaf protein. Protein was extracted according to Matai *et al*., (1971) and leaf protein was precipitated by heating the whole green juice to 80°C. Extractability percentage was calculated following Buyers and Sturrock (1965). Data on fresh yield and extractability of the plants are presented in Table 21.4.

Sixteen plants had fresh yield values above 1000 g per m^2, emergent plants being the most productive. As certain species and inaccessible parts of the sampled species were omitted by the sampling method, total yield of the plants could not be obtained. Extractability is a very important aspect since the amount of leaf protein that is ultimately obtained depends directly on the extractability of the plant. Thirteen plants exhibited protein extractability above 20% which is low when compared to other crop plants (Byers and Sturrock, 1965). Selection of promising species from the thirty plants studied was made on the basis of leaf protein with a nitrogen content of 6% and more, prepared from plants with more than 20% protein nitrogen extractability. The chemical composition of leaf protein prepared from selected plants is given in Table 21.5.

Leaf protein extracted from six aquatic weeds had more than 50% crude protein. All leaf proteins were a good source of B-carotene which is nutritionally important in view of the widespread prevalence of Vitamin A deficiency leading to xeropthalmia and blindness. *In vitro* digestibility, useful for rapid screening of nutritional quality of any protein, was measured by preliminary hydrolysis with pepsin followed by pancreatin

Table 21.4 : Fresh yield and extractability data of aquatic plants

Plants	Fresh yield	Pulp DM	Pulp N	Fibre DM	Fibre N	Extractability: Total N	Extractability: Protein N	N in dry LP
	g per m²	%	%	%	%	%	%	%
FLOATING								
Azolla pinnata	963	7.83	3.42	17.00	3.06	28.18	21.23	9.36
Eichhornia crassipes	2507	9.00	1.22	19.67	1.55	16.65	6.14	6.57
Gentiana lutea	947	6.50	2.76	12.33	2.70	27.90	11.96	5.26
Ipomoea reptans	997	5.17	3.06	10.83	2.92	48.11	33.84	8.40
Jussiaea repens	903	7.50	3.86	21.07	3.55	24.02	9.57	5.10
Lemna minor	1383	6.67	2.45	15.50	2.49	32.65	21.49	6.16
Limnanthemum critatum	1493	5.03	3.56	11.90	2.77	50.70	39.43	8.71
Nelumbo nucifera	nd	11.00	3.21	19.00	3.09	10.17	4.32	5.92
Nymphaea nouchali	12.53	3.00	2.20	13.50	2.32	35.28	21.17	4.58
Nymphae stellata	1680	8.17	2.90	17.00	2.62	21.80	12.61	5.56
Pistia stratiotes	nd	5.50	2.52	129.00	2.66	26.64	11.53	5.37
Salvinia auriculata	1660	5.17	2.44	17.33	1.95	41.31	27.00	8.45
Trapa bispinosa	nd	10.83	2.04	20.67	2.21	17.31	9.33	2.58
SUBMERGED								
Ceratophyllum demersum	nd	4.83	2.74	15.67	2.70	36.74	19.78	3.94
Hydrilla verticillata	nd	7.00	2.78	16.33	3.29	26.71	10.69	5.17
Ottellia alismoides	1207	6.67	2.94	23.17	1.71	44.98	30.77	6.84
Vallisneria spiralis	nd	5.50	3.15	8.33	2.79	15.75	5.72	5.44
EMERGENT								
Allmania nodiflora	1590	10.50	3.11	18.83	2.27	44.22	33.91	10.03
Alternanthera philoxeroides	1323	11.83	2.57	20.33	2.46	31.36	22.91	7.76
Commelina benghalensis	1160	6.83	2.96	20.50	2.55	31.86	17.54	5.19
Enhydra fluctuans	nd	8.00	3.06	16.83	3.01	26.90	7.55	3.06
Hygrophila spinosa	1017	9.33	3.19	17.67	3.05	36.57	20.40	7.16
Hydrocotle asiatica	567	6.00	2.03	18.33	1.86	25.71	21.03	5.78
Linophila heterophylla	1040	8.00	2.22	24.00	2.03	24.83	9.25	3.98
Lippia geminata	1937	12.67	2.61	22.50	2.97	18.91	5.41	5.42
Marsilea quadrifolia	590	7.17	3.25	13.00	2.96	20.34	3.21	3.44
Mikania scandens	1123	7.33	3.00	20.83	2.42	41.07	22.46	8.14
Phrogmites karka	2310	14.67	1.79	20.00	1.99	15.91	11.11	6.72
Polygonum barbatum	1160	12.67	3.21	27.17	3.29	30.67	23.52	6.62
Typha latifolia	2650	17.00	1.49	26.67	1.62	3.57	0.61	4.85

DM = dry matter, N = nitrogen; LP = leaf protein; nd = not determined

(Saunders *et al.*, 1973). Eight of the eleven leaf proteins had digestibility above 54%.

The final evaluation of leaf protein as a source of dietary protein was done by actual feeding experiments where the ability of leaf protein to supplement cereal based diets was studied. As per availability, three aquatic weeds, namely, *Alternanthera philoxeroides, Pistia stratiotes* and *Lemna minor* were used for large scale extraction of leaf protein. The protein quality of leaf proteins was estimated by rat feeding trials using the Protein Efficiency Ratio (PER) method (AOAC, 1984). The five diets comprised a standard caesin diet, a wheat flour diet and three diets and leaf protein supplemented to wheat flour in 3:7 ratio. The caesin and test diets were prepared by incorporating each material separately at the expense of starch in the basal protein free diet to give 10% protein. The observed mean body weight of experimental rats in various groups, the quantity of food consumed and PER values are reported in Table 21.6 along with their dietary protein contents. All values are mean of eight observations.

Table 21.5: Proximate composition, x-carotene and *in vitro* digestibility (% dry weight) of selected leaf proteins

Leaf protein	*Crude protein %*	*Crude fat %*	*Crude fibre %*	*Ash %*	*B carotene ug/g*	*In-Vitro digestibility %*
FLOATING						
Allmania nodiflora	62.7	11.1	2.7	9.0	782.4	69.8
Altermanthera philoxeroides	48.5	15.2	3.8	8.1	408.6	55.7
Azolla pinnata	58.5	17.5	2.6	4.8	671.7	63.3
Hygrophila spinosa	44.8	19.8	4.2	10.1	453.1	42.1
Ipomoea reptans	52.4	17.1	1.9	4.7	848.7	54.1
Lemina minor	32.9	11.3	2.3	12.2	626.5	59.2
Limnanthemum critatum	54.4	16.0	3.6	5.8	659.1	64.3
Milkania scandens	50.9	19.0	3.5	4.8	607.3	60.9
Ottelia alismoides	42.8	17.2	2.0	8.8	580.6	46.8
Pistia stratiotes	53.2	16.2	1.6	9.9	726.2	68.5
Polygonum barbatum	41.4	18.2	2.3	11.6	595.2	44.8

The PER of the wheat flour group was very low when compared to the caesin fed group, but with the addition of leaf protein it improved in all the three supplemented groups. Protein extracted from *Pistia stratiotes* and *Lemna minor* showed better response when supplemented to wheat flour in comparison to that of *Alternanthera philoxeroides*. The presence of undesirable flavouring constituents in *Alternanthera philoxeroides* leaf protein probably resulted in low acceptability of this diet which is reflected in the diet intake of the rats in this group. It is evident that leaf protein prepared from *Pistia stratiotes* and *Lemna minor* at low levels of supplementation greatly improved the poor quality of a wheat diet. Moreover, fresh yield of both weeds was more than 1000 g per m^2 (see Table 21.4).

Thus, some aquatic plants provide an abundant natural resource which could be used in many localities to alleviate nutritional inadequacies by supplementing traditional foods and feeds.

Table 21.6 : Dietary protein content, food intake, body weight gain and protein efficiency ratio of rats fed various diets[a].

Groups	*Dietary protein (g)*	*Diet intake (g)*	*Gain in weight (g)*	*Protein efficiency ratio (PER)*
1. Caesin	10.1	309.0	74.8	2.40
2. Wheat flour	10.4	267.1	26.4	0.95
3. Wheat flour + *Pistia* leaf protein	10.2	295.9	46.9	1.55
4. Wheat flour + *Altermanthera* leaf protein	10.0	212.9	24.1	1.14
5. Wheat flour + *Lemna* leaf protein	10.1	304.5	44.1	1.43

[a] Source : Dewanji *et al.*, 1990

References

Aboaba, F.O., (1973) Management of aquatic vegetation. World Crops. V. 25, pp. 21-24.

Aitchinson, E., (1988) Cereal straw and stubble for sheep feed J. Agric. V. 29, pp. 96-101.

AOAC., (1984) Official Methods of Analysis (Horwitz, W. ed.), Washington, D.C. 1141 pp.

Banerjee, A. and Matai, S., (1990) Composition of Indian aquatic plants in relation of utilization as animal forage. J. Aquat. Plant Manage. V. 28, pp. 69-73.

Banerjee, G.C., (1988) Feeds and principles of animal nutrition, Oxford and IBH Publishing Co., New Delhi, 636 pp.

Bondi. A, and Alumot, E., (1987) Anti-nutritive factors in animal feedstuffs and their effects on livestock. Prog. Food Nutr. Sci. V. 11, pp. 115-51.

Byers, M. and Sturrock, J.W., (1965) The yields of leaf protein extracted by large scale processing of various crops. J. Sci. Food Agric. V. 16, pp. 341-55.

Dewanji, A. Matai, S and Chakravarty, I., (1990) Studies on quality of leaf proteins prepared from aquatic weeds. In : Proceedings EWRS 8th Symposium on Aquatic Weeds (Barrett, P.R.F., Greaves, M.P., Murphy, K.J., Pieterse, A.H., Wade, P.M. and Wallsten, M. eds.), Uppsala, Sweden. pp. 85-90.

Dolberg, F., Saadullah, M. and Haque, M., (1981) A short review of the feeding value of water plants. Trop. Anim. Prod. V. 6, pp. 322-26.

Hultin, E. and Torsell, K., (1965) Akaloid screening of Swedish plants, Phytochem. V. 4, pp. 425-33.

Majumdar, N.C., (1965) Aquatic and semi-aquatic flora of Calcutta and adjacent localities. Bull. Botan. Soc. Bengal. V. 19, pp. 10-17.

Matai, S. Bagchi, D.K. and Roychowdhury, S., (1971) Leaf protein from some plants in West Bengal. Sci. Engg. V. 24, pp. 102-05.

McClure, J.W., (1970) Secondary constituents of aquatic angiosperms. In : Phytochemical Phylogy (Harborne, J.B. ed.), Academic Press, pp. 233-68.

NAS., (1984) Making aquatic weeds useful : some perspectives for developing countries. National Academy of Sciences, Washington D.C., 175 pp.

Pirie, N.W., (1987) Leaf protein and its-products in human and animal nutrition. Cambridge University Press, 209 pp.

Saunders, R.M. Connor, M.A. Booth, A.N. Bickoff, E.M. and Kohler, G.O., (1973) Measurement of digestibility of alfalfa concentrates by *in vivo* and *in vitro* methods. J. Nutr. V. 103, pp. 530-35.

Varshney, C.K. and Singh, K.P., (1976) A survey of the aquatic weed problems in India. In : Aquatic weeds in south east Asia (Varshney, C.K. and Rzoska, J. eds.) W. Junk, the Hague, pp. 31-41.

Wright, M.J. and Davison, K.L., (1964) Nitrate accumulation in crops and nitrate poisoning in animals. In : Advances in Agronomy (Norman, A.G. ed.), Academic Press, V. 16, pp. 197-247.

22

Ecology of Fishes in Wetland Ecosystems of Kashmir with Emphasis in on their Management

ASHOK. K. PANDIT

Introduction

Nature is very rich where water meets land. Ecologically wetlands are viewed to be among the most productive ecosystems on the earth. The high primary productivity of wetlands has been demonstrated to be associated with high secondary production and more particularly the fish production (de la Cruz, 1973; Lieth and Whittaker, 1975; Turner, 1977). This depends on the production of detritus and its flux to adjacent water bodies where it is incorporated into the food web. It is estimated that two third of the world fisheries depend on the fish that have their breeding or nursing grounds in wetlands (Braakhekee and Marchand, 1987). In India as much as 80% of the catch of commercial fisheries in the Ganges-Brahmaputra estuary is based on the Sunderbans wetland complex (Malthy, 1986). In view of the high protein yield of wetlands in terms of fish production, it is imperative to have a fair understanding of the ecology of fishes in diverse wetlands of Kashmir, differing in water depth, vegetation, food supply, trophic status, etc.

Study Sites

The wetlands, selected for the present study are situated in the flood-plains of River Jhelum (Fig. 22.1) and classified by Pandit and Fotedar (1982) into four different types :

(a) Temporary and Shallow (Nowgam)

(b) Temporary and Deep (Mirgund)

(c) Permanent and Shallow (Haigam and Malgam)

(d) Permanent and Deep (Hokarsar)

The location, area and detailed description of these five wetlands is published elsewhere (Kaul and Pandit, 1981; Pandit 1980, 1991; Pandit and Fotedar, 1982; Pandit and Kaul, 1981). The distribution and biomass production of macrophytes growing in the wetlands sites have been discussed by Pandit (1980, 1991) and Kaul (1982).

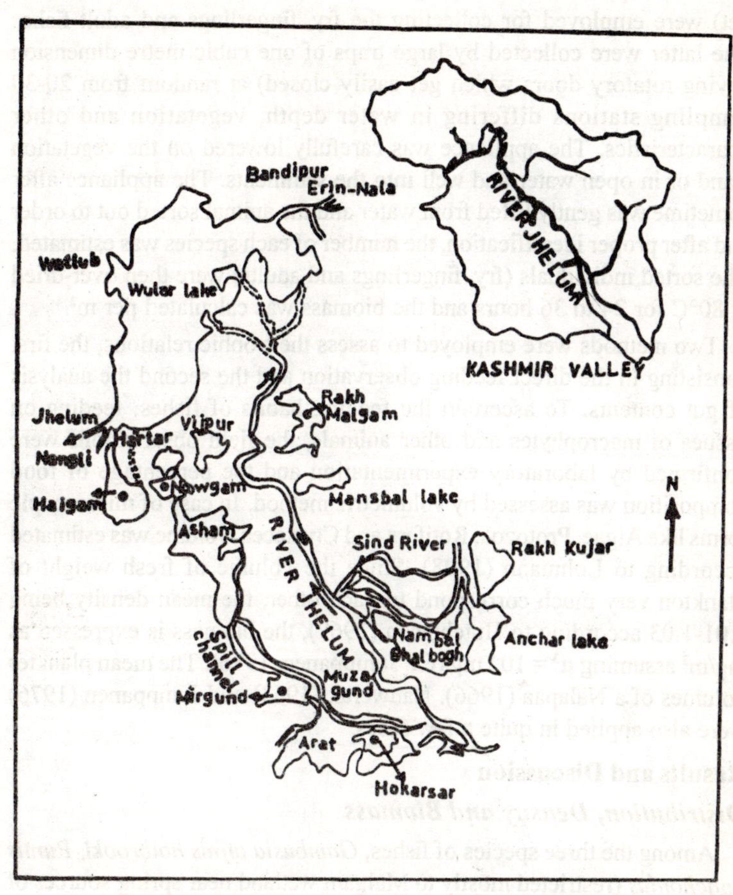

Fig. 22.1 : Location of wetlands in Kashmir

Methods of Study

Cubic traps of different sizes (metal frames covered with close meshed net) were employed for collecting the fry, fingerlings and adult fishes (the latter were collected by large traps of one cubic metre dimension having rotatory doors which get easily closed) at random from 20-30 sampling stations differing in water depth, vegetation and other characteristics. The appliance was carefully lowered on the vegetation stand or in open water and well into the sediments. The appliance after sometime was gently lifted from water and the animal sorted out to order and after proper identification, the number of each species was estimated. The sorted individuals (fry, fingerlings and adults) were then over-dried at 80°C for 24 to 36 hours and the biomass was calculated per m^2.

Two methods were employed to assess the trophic relations, the first consisting in the direct feeding observation and the second the analysis of gut contents. To ascertain the feeding habits of fishes, feeding on tissues of macrophytes and other animals, the field observations were confirmed by laboratory experimentation and the percentage of food composition was assessed by volumetric method. In case of microscopic forms like Algae, Protozoa, Rotifera and Crustacea, volume was estimated according to Lohmann (1908). Since the volume of fresh weight of plankton very much correspond to each other, the mean density being 1.01-1.03 according to Hutchinson (1967), the biomass is expressed as mg/m^3 assuming $\mu^3 = 10^{-9}$ mg (c.f. Jumppanen, 1976). The mean plankter volumes of a Nalapaa (1966). Nauwerak (1963) and Jumppanen (1976) were also applied in quite a few cases.

Results and Discussion

Distribution, Density and Biomass

Among the three species of fishes, *Gambusia affnis holbrooki, Puntis conchonius* (restricted mostly to Malgam wetland near spring sources of water) and *Cyprinus carpio*, the last one is consumed by human beings. *Cyprinus carpio communis*, constituted the main fish type in Nowgam, Haigam, Mirgund and Hokarsar wetlands while the mirror carp (*Cyprinus carpio specularis*) seems to be more abundant in Malgam and Horarsar only.

The seasonal pattern in both the density and biomass varied considerably the former tracing a bimodal pattern with spring and autumn peaks and the latter a unimodal pattern with the only autumn maximum. This is because that both the sub species of exotic carp breed twice a year, although a larger number of fry are produced in late spring as compared to autumn. The maintenance of large fish population in their fry stage is due to better food supply provided by plankton population and that of adult carp by large amount of organic detritus resulting from the decay of macrophytes and the transport of materials via tributary inputs. The same very reason explains the maximum biomass of fishes in autumn. The peak density values of 135-844 in div. m^{-2} during spring are mainly due to the presence of large populations of fry and fingerlings while the maximum biomass of 20.00-56.34 gm^{-2} obtained in fall is due to the bigger specimens of the fish in the samples. The density (maximum 32.11 indiv. m^{-2}) is, however, modest in summer as opposed to winter (0.14-12.02 indiv. m^{-2}) (Table 22.1).

A comparison of fish biomass in five different wetlands of Kashmir revealed Mirgund to be most productive and Nowgam least productive (Table 22.2). The increased fish biomass in wetlands like Mirgund and Malgam compared to macrophytic biomass in these biotopes is due to the high proportion of the edible content in the organic detritus as indicated by the gut analysis of these fishes as well as the other studies (Pandit, 1980; Pandit and Kaul, 1981).

Trophic Relations

The trophic relations of fishes in wetlands of Kashmir are very complex (Fig. 22.2) as they feed on a variety of plants, detritus and small animals and are in turn eaten by large predators. The fishes, therefore, form important links in the food-chain of herons, kites, vultures, king fishers and, of course, the occasional visitors like dogs particularly under drought conditions. The diversity of fishes coupled with the variety of potential foods would suggest very complex and variable nature of trophic relations as delineated in the food web of wetlands (Pandit, 1980; Pandit and Kaul, 1981; Pandit and Fotedar, 1982; Kaul *et al.*, 1980). However, the annual food spectrum of different fish species is given in Table 22.3.

(i) *Cyprinus carpio*

(a) New Hatching Fry

The first natural food components of carp fry on the third day after hatching were Crustacea and Rotifera in one of the spawning grounds, small chironmid larvae in the second and Rotifera, Crustacea and Protozoa in the third, though in majority of the cases they feed on zooplankton. The proportions of algae were quite insignificant as compared to animal food items contributing 98.3 % of the total food volume. However, none of the animal groups was dominant or decisive in the food, indicating the association of this stage of fish to a wide variety of feeding niches

Table 22.1 : Seasonal variation in the population denisty (D-indiv. m^{-2}) and biomass (B-g m^{-2}) of fishes in five different wetlands to Kashmir.

Species	*Spring*		*Summer*		*Autumn*		*Winter*	
	D	*B*	*D*	*B*	*D*	*B*	*D*	*B*
I. NOWGAM	ind./m^2	g/m^2	ind./m^2	g/m^2	ind./m^2	g/m^2	ind./m^2	g/m^2
Cyprinus carpio (F1)	133.00	3.268	-	-	152.00	11.552	-	-
Cyprinus carpio (F2)	-	-	3.00	9.954	6.00	4.272	-	-
C. c. specularis (a)	-	-	-	-	-	-	-	-
C.c. communis (a)	0.08	1.750	0.50	6.400	0.15	3.281	0.02	0.440
Puntius conchonius	-	-	-	-	-	-	-	-
Gambusia affinis	15.00	1.140	30.00	1.800	12.00	0.912	5.00	0.72
Total	148.08	5.158	33.50	18.154	170.15	20.017	5.02	0.512
II. MALGAM								
Cyprinus carpio (F1)	125.00	0.750	13.00	0.270	86.00	4.265	-	-
Cyprinus carpio (F2)	-	-	-	-	4.00	13.272	-	-
C. c. specularis (a)	0.20	10.500	0.25	16.800	0.16	4.750	0.02	4.500
C.c. communis (a)	-	-	-	-	-	-	-	-
Puntius conchonius	-	-	-	-	30.00	13.146	-	-
Gambusia affinis	43.00	6.106	98.00	1.911	18.00	0.893	12.00	0.657
Total	168.20	17.3567	111.25	18.981	138.16	36.326	12.02	5.157
III. HAIGAM								
Cyprinus carpio (F1)	819.00	5.816	50.00	2.475	382.00	22.920	-	-
Cyprinus carpio (F2)	-	-	0.50	1.070	9.00	4.005	-	-
C. c. specularis (a)	-	-	-	-	-	-	-	-
C.c. communis (a)	-	-	0.45	8.125	0.20	3.750	0.20	5.650
Puntius conchonius	-	-	-	-	-	-	-	-
Gambusia affinis	33.00	1.129	52.00	11.544	225.00	11.383	2.80	0.159
Total	852.00	6.945	102.95	23.214	616.20	42.058	3.00	5.809
IV. MIRGUND								
Cyprinus carpio (F1)	301.00	2.137	5.00	0.070	491.00	18.903	-	-
Cyprinus carpio (F2)	-	-	-	-	6.00	13.506	-	-
C. c. specularis (a)	-	-	-	-	-	-	-	-
C.c. communis (a)	0.15	12.000	0.20	14.444	0.60	24.000	0.14	5.000
Puntius conchonius	-	-	-	-	-	-	-	-
Gambusia affinis	19.00	0.712	32.00	1.200	6.00	0.225	-	-
Total	320.15	14.849	37.20	15.714	503.60	56.634	0.14	5.000

Contd.

Table 22.1 (Contd.)

Species	Spring D	Spring B	Summer D	Summer B	Autumn D	Autumn B	Winter D	Winter B
V. HOKARSAR								
Cyprinus carpio (F1)	87.00	2.465	-	-	78.00	2.605	-	-
Cyprinus carpio (F2)	-	-	2.50	5.348	6.00	12.837	4.00	2.558
C. c. specularis (a)	0.08	6.000	0.16	10.000	0.16	18.888	-	-
C.c. communis (a)	0.01	0.575	0.10	6.250	0.50	6.100	0.05	1.710
Puntius conchonius	-	-	-	-	-	-	-	-
Gambusia affinis	48.00	2.772	29.00	4.451	5.00	0.288	1.00	0.057
Total	135.09	11.812	31.76	26.019	89.66	40.718	5.05	4.325

F1 = Fry F2 = Fingerlings a = Adult

Table 22.2 : Comparison of five different wetlands on the basis of fish biomass* (g m^{-2}) and macrophytic production (g m^{-2})**

Wetlands	Macrophytic ANP	BNP	TNP	Production L	(g/m^2) AGB (Ranges)	Fish Biomass (g/m^2) (Ranges)
Nowgam	648.66	260.16	908.82	121.00	64.00-648.66	0.51-20.02
Malgam	239.25	93.19	332.44	173.00	20.64-239.25	5.16-36.33
Haigam	1445.91	615.06	2060.97	167.00	174.92-1445.91	5.81-42.06
Mirgund	201.18	1.92	203.10	76.00	4.62-201.18	5.00-56.63
Hokarsar	674.33	72.52	746.85	137.00	17.08-674.33	4.33-40.73

ANP - Above ground Net Annual Production
BNP - Below-ground Net Annual Production
TNP - Total Net Annual Production
L - Litter Production
AGB - Above Ground Biomass

* Pandit (1980)
** S. Kaul (1982)

and as such protozoans, rotifers, cladocerans, copepods and insects made up moderate proportion of 17.5, 20.6, 23.2, 15.4 and 19.3% respectively. Ostracod, however, were poorly represented (2.3%).

(b) Fry (3-5 cm)

Carp fry is an omnivore with preference for zooplankton. Under natural conditions, algae form the bulk of food and account as much as 59% of the total food consumed. Algae (*Navicula* spp. *Cymbella* spp., *Pinnularia* spp., *Gomphonema parvulum, G. constrictum, G. venustum, Synedra*

Table 22.3 : Percentage Composition (by volume) of various food components in the guts of fish species inhabiting the five wetlands of Kashmir (1976-79).

Food items	*Cyprinus carpio*				*Gambusia affinis*	*Puntius conchonius*
	NHF	*F1*	*F2*	*A*	*A*	*A*
PLANTS						
Cyanophyceae	-	6.4	7.0	7.3	-	6.1
Chlorophyceae	1.4	39.0	32.1	29.4	-	46.7
Bacillariophyceae	0.2	7.3	15.0	12.2	-	39.3
Eugleninae	0.1	4.2	4.1	4.0	-	2.3
Other algae	-	1.1	0.7	0.6	-	0.6
Total algae	1.7	58.0	58.9	53.5	-	95.0
Macrophytes (Dead)	-	0.9	5.0	19.2	-	3.1
Macrophytes (Living)	-	0.6	2.6	7.9	-	0.9
Fungi	-	-	-	0.2	-	-
Total Plant Material	1.7	59.5	66.5	80.8	-	99.0
ANIMALS						
Protozoa	17.5	4.2	3.4	1.5	2.2	-
Rotifera	20.6	6.4	5.2	0.6	1.6	-
Cladocera	23.2	13.0	11.5	1.2	13.0	-
Copepoda	15.4	7.4	6.2	0.9	9.7	-
Ostracoda	2.3	2.5	-	-	1.3	-
Micro-crustacea	40.9	22.9	17.7	2.1	24.0	-
Total zooplankton	79.0	33.5	26.3	4.2	27.7	-
Macro-crustacea	-	-	-	-	0.6	-
Arachnida	-	-	-	-	1.5	-
Insecta	19.3	6.5	4.1	2.1	61.4	-
Nemathelminths	-	0.5	2.1	-	0.2	-
Annelida	-	-	0.4	1.3	2.6	0
Mollusca	-	-	-	2.0	6.0	-
Total Animal Material	98.3	40.5	32.9	9.6	100.0	
Detritus	-	-	0.6	9.1	-	1.0
Calcium deposits	-	-	0.5	-	-	-

NHF - Newly Hatched Fry

F1 - Fry (3-5 cm)

F2 - Fingerlings

A - Adult

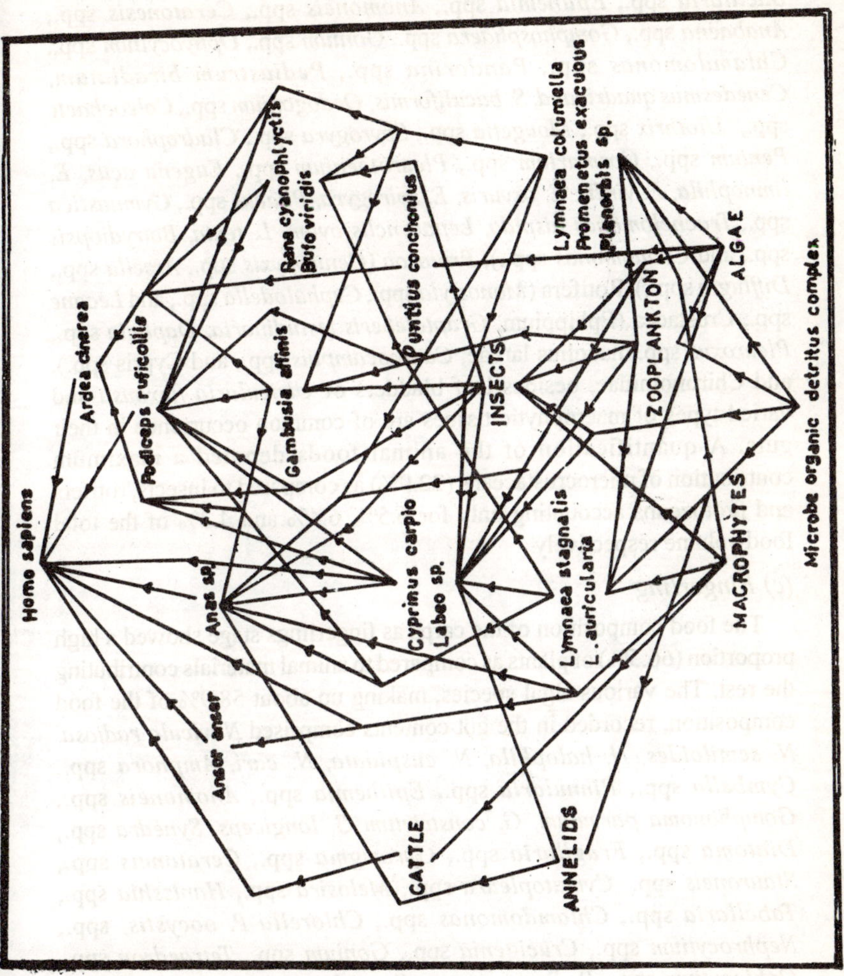

Fig. 22.2 : A typical food web of wetlands in Kashmir, showing the tropic relations of fishes.

spp., *Diatoma* spp., *Fragilaria* spp., *Gyrosigma* spp., *Melosira* spp., *Bacillaria* spp., *Epithemia* spp., *Anomoneis* spp., *Ceratonesis* spp., *Anabaena* spp., *Gomphosphaera* spp., *Gonium* spp., *Ophyocytium* spp., *Chlamdomonas* spp., *Pandorina* spp., *Pediastrum biradiatum, Cenedesmus quadricaud, S. baculiformis, Oedogonium* spp., *Coleochaete* spp., *Ulothrix* spp., *Mougetia* spp., *Spyrogyra* spp., *Cladrophora* spp., *Penium* spp., *Cosmarium* spp., *Pleurotaenium* spp., *Eugena acus, E. limnophila, E. virdis. E. oxyuris, E. spirogyra, Phacus* spp., *Gymnastica* spp., *Trachelomonas hispida, Lepocynclis ovum, L. texta, Botrydiopsis* spp., and *Cyanomonas* spp.,), *Protozoa* (*Centropyxis* spp., *Arcella* spp., *Difflugia* spp.,), Rotifera (*Monostyla* spp., *Cephalodella* spp., and *Lecane* spp., Crustacea (Ephippium, *Graptoleberis testudinaria, Daphnia* spp., *Pleuroxus* spp., nauplius larvae, *Canthocamptus* spp., and Cypris spp.), and Chironomidae, besides leaf bladders of *Utricularia flexousa* and varied types of macrophytic tissues are of common occurrence in their guts. A quantification of the animal foods denoted a maximum contribution of microcrustaceans (22.9%) as compared to insects, rotifers and protozoans accounting only for 6.5%, 6.4% and 4.2% of the total food volume respectively.

(c) Fingerling

The food composition of the carps as fingerlings stage showed a high proportion (66.5%) of plants as compared to animal materials contributing the rest. The various algal species, making up about 58.9% of the food composition, recorded in the gut contents comprised *Navicula radiosa, N. semiloides, N. halophila, N. cuspidata, N. cari, Amphora* spp., *Cymbella* spp., *Pinnularia* spp., *Epithemia* spp., *Anomoneis* spp., *Gomphonoma parvulum, G. constrictum G. longiceps, Synedra* spp., *Diatoma* spp., *Fragilaria* spp., *Gyrosigma* spp., *Ceratoneis* spp., *Stauroneis* spp., *Cymatopleura* spp., *Melosira* spp., *Hantzshia* spp., *Tabellaria* spp., *Chlamdomonas* spp., *Chlorella P. oocystis,* spp., *Nephrocytium* spp., *Crucigenia* spp., *Gonium* spp., *Tetraedron* spp., *Ophiocytium* spp., *Pediastrum* spp., *Scenedesmus* spp., *Ultohrix* spp., *Oedogonium* spp., *Mougetia* spp., *Gleonodinium* spp., *Anabaena* spp., *Oscillitoria* spp., *Spirulina* spp., *Gomphosphaera* spp., *Chroococcus* spp., *Synechococcus* spp., *Merismopedia* spp., *Euglena acsu, E. limnophila, E. virdis, Astatia* spp., *Phacus longicauda, P. caudatus* and *P. platalea.* The percentage contribution of Protozoa (*Centropyxiz ecornis, C. constrictum, Lesquereusia modesta, Arcella* spp., *Amobea* spp.), towards the total animal diet decreased to only 3.4% from that of fry stage. A similar decrease was recorded for rotifers (*Keratella* spp., *Brachionus* spp., *Filinia* spp., *Lecane* spp., *Monostyla* spp., *Cephalodella* spp.), crustaceans (Epihippium, *Macrothrix* spp., *Pseudosida bidentata, Alona* spp., *Bosmina* spp., *Daphnia* spp., *nauplius* larvae, *Cyclops* spp., *Macrocyclops* spp., *Diaptomus* spp., *Canthocamptus* spp., and *Bryocamptus* spp.) and insects like chironomids and others, the three groups contributing only 5.2%, 17.7% and 4.1% respectively. However, both annelids and Nematheminthes showed proportions of 0.4% and 2.1% respectively. Comparing the food composition of fingerlings with that of fry stages, the proportions of Chlorophyceae and animal materials seemed to decrease against an increase in diatoms, blue-greens and macrophytes.

(d) Adult carp

The feeding habits of the fish are those of a grubber as it continuously protrudes its lips and forms a suction funnel through which it imbibes pedon or decaying organic debris sedimenting at the bottom (Subla, 1970). Being primarily a bottom feeding detritivore or scavenger, *Cyprinus* roams about in search of food in all shallow regions, being at home even in highly eutrophicated sewage outfalls and side channels. The common hardy carp, though an omnivore yet primarily a herbivore, feeds on a wide variety of algae which put together contribute as much as 53.5% of the total food volume against 27.1% contribution of higher plants. The fish species is a roiler and uproots submerged vegetations as it scavenges around in bottom while looking for insects and other small animal life, a fact well corroborated by the studies of Subla (1970). The striking peculiarity in the food composition of *Cyprinus carpio* is the high amount of detritus (9.1%) as well as calcium deposits (0.5%), besides negligible quantities of fungi. The overall plant life contributing 80.8% of the annual feed consists of macrophytes like *Ceratophyllum demersum, Nymphoides peltata, Potamogeton natans, P. crispus, P. lucens, Hydrocharis morus-ranae, Hydrilla verticillata, Lemna* spp., *Salvinia natans, Myriophyllum spicatum, Utricularia flexuosa* and *Chara* spp., and a multitudinous number of algal forms represented by Cyanophyceae (*Gomphosphaera* spp., *Microcystis* spp., *Anabaena* spp., and *Rivularia* spp.), Chlorophyceae (*Chalmdomonas* spp., *Spirogyra* spp., *Oedogonium* spp., *Mougetia* spp., *Ulothrix* spp., *Zygnema* spp., *Cladophora* spp., and *Cosmarium* spp.), Bacillariophyceae (*Navicula* spp., *Cymbella* spp., and *Pinnulara* spp., *Synedra* spp., *Diatoma* spp., *Fragilaria* spp., *Epithemia* spp., *Anomoneis* spp., *Ceratoneis* spp., *Gomphonema* spp., *Coconeis* spp., *Tabellaria* spp., *Gyrosigma* spp., *Suirella* spp., *Nitzschia* spp., *Cyclotella* spp., *Diatomella* spp., and *Bacillaria* spp.) and Eugleninae

(*Euglena* spp., *Trachelomonas hispida, T. volcocina, Lepocynclis texta* and *Phacus* spp.,). The remaining 9.6 % of the animal food material, on which it feeds selectively, comprised 1.5% protozoans, (*Pareuglypha reticulata, Arcella* spp., *Difflugia* sp., *Wailsella* sp., and *Centropyxis* sp.), 0.8% rotifers (*Cephalodella* sp., *Monostyla* spp., *Keratella* spp., *Epiphanes* spp., *Synchaeta* spp., *Brachionus* spp., and *Asplanchna* spp.), 2.1% crustacens (Epihippium, *Daphnia* spp., *Bosmina* spp., *Chydorus* spp., *Alona* spp., *Cyclops* spp. and *Canthocamptus* spp.), 1.3% annelids, 2% molluscs and 2.1% chironomids and other insects. Thus, a purview of the feeding ecology of the fish would, at the outset, reveal an increasing proportion of macrophytes and detritus against the decreasing proportion of both algae and animal food with age.

(ii) *Gambusia affnis*

The small mosquito fish, though largely regarded to be an insectivore, preys upon highly varied animal life comprising Protozoa (*Arcella mitrata, A. vulgaris, A. megastoma, Centropyxiz ecornis, C. arcelloides, Euglypha* spp., *Pareuglypha* spp., *Difflugia lebes* and *Amoeba* spp.), Rotifera (*Platyias* spp., *Monostyla* spp., *Synchaeta pectinata, Lecane ohioensis, Mytilina mucronata, Keratella valga, K. cochlearis* and *Brachionus* spp.), Crustacea (*Graptolebris testudinaria, Monospilus dispar, Bosmina* spp., *Daphina* spp., *Ceriodaphnia reticulata, Moina affinis, Scapholeberis kingi, Simocephalus* spp., *Macrothrix* spp., *Acroperus harpae, Pleuroxus denticulatus, Eurycerus* spp., *Camptocercus* spp., *Leydigia* spp., *Alona intermedia, A. affinis, Alonella exigua, Chydorus ovalis, C. gibbus, C. barroisi, C. sphericus, Epihippium,* nauplis larvae, *Cyclops vernalis, C. venustrum, C. bicuspidatus, Cyclops* spp., *Macrocyclops* spp., *Eucyclops agilis, Bryocamptus* spp., *Canthocamptus* spp., *Cypris* spp., *Eucypris* spp,, and *Gammarus* spp.), Arachnida (Aranae and Acari), Insecta (Chironomidae, Culicidae, dixid larvae, Rhagonoidae, Coleoptera larvae, Hydraenidae, Hymenoptera, *Coraxia* spp., Anisoptera and Zygoptera nymphs), Nematoda, Tubificidae, Mollusca (*Lymnaea* spp., *Promenetus exacuous* and *Planorbis* spp.) and fry of *Cyprinus carpio.* The food composition of the fish in terms of percentage contribution revealed fairly high proportion of insects (61.4%), the other animals in order of their importance as food items being micro-crustaceans, molluscs, annelids, protozoans, rotifers, arachnids, macro-crustaceans and nematodes with % values of 2.4%, 6%, 2.6%, 2.2%, 1.5%, 1.5%, 0.8% and 0.2% respectively.

(iii) *Puntius conchonius*

An analysis of the gut contents reveals the fish to be 99% herbivore as the algae contributes as much as 95% against the remaining 5% only being contributed by macrophytes (4%) and detritus (1%). Among the algal constituents, the proportion of chlorophycean forms like *Ulothrix* spp., *Oedogonium* spp., and *Spirogyra* spp., was as much as 46.7% to be followed closely by Bacillariophyceae (39.3% and represented mainly by *Navicula* spp., *Cymbella* spp., *Gomphonema* spp., *Synedra* spp., *Diatoma* spp., *Fragilaria* spp., *Coconeis* spp., *Surirella* spp., and *Bacillaria* spp.,) and Cyanophyceae (6.1% and represented mainly by *Microcystis* spp., *Lyngbya* spp., *Oscillatoria* spp., *Anabaena* spp., *Rivularia* spp., *Gomphophaera* spp.,) against the poor representation of 2.3 and 0.6% for Eugleninae (*Trachelomonas* spp.) and other insignificant groups respectively.

Resource Utilization

Man in the neighbouring areas of wetlands is dependent on the fish resources for its food and is, therefore, directly related to ecosystems upon which he has profound influence. This is largely due to the availability of high yield of fishes within these biotops. It has been observed that the nutritional status of Kashmir fishes is quite high as is evident from their proximate analyses (on dry wt. basis) revealing 57.2-62.8% protein, 10.6-14.4% lipids, 13.8-22.1% ash (minerals) and 6.4-12.4 carbohydrates (Table 22.4). Among the piscine forms, the small mosquito fish (*Gambusia affinis*) gained superiority over other two types in so far as the high proportion of protein is concerned, but the species cannot be exploited as a food source because of its very small size (Pandit and Qadri, 1986). Levels of crude protein in the freshwater fishes are reported to be slightly lower than the values estimated for such marine fishes as India mackerel *Rastrelliger kanagurta,* (Sawant and Magar, 1961).

There is no information available about the total annual fish catch and the depletion of fish resources in wetlands but definitely during drought which mostly occurs in autumn the wetlands are subjected to great scarcity of water as a consequence of which the benthic fauna, fish and other animals are exposed to vulnerable predation by dogs, kites, vultures, mongoose (more particularly in Malgam wetland) and other carnivorous animals bringing huge catastrophy to the fish source of temporarily drying wetlands like Nowgam and Mirgund. There are speculations that snow trout (*Schizothorax* spp.) existed in these

Table 22.4 : Proximate analyses (Dry-weight basis) of fishes of Kashmir (Pandit and Qadri, 1986).

Species	% *Crude Protein*	% *Lipids*	% *Ash*	% *Non-nitrogen-ous substances*
Cyprinus carpio	57.20	14.36	22.05	6.39
Puntius conchonius	60.63	10.59	20.25	8.53
Gambusia affinis	62.76	11.12	13.77	12.35

* Carbohydrates including fibre

ecosystems before 1950 but has now been completely eliminated from these biotopes due to their shallow nature coupled with high trophic status as compared to lakes where it still makes 30% of the total fish population (Kaul and Pandit, 1979).

Fish Management

The traditional attitude that wetlands are "wastelands" serving as sinks for floods or dumping grounds for waste has over the years led to the deterioration of wetlands and depletion of their natural resources. As a consequent thereof, the wetlands have been drained, filled and reclaimed for agricultural purposes, road, building and, housing developments. These operations, destroying the wetland character of the land, are partly a manifestation of fast urbanisation and partly an outcome of the campaign "Grow more Food" in 1950. Multiple activities of man by way of cattle grazing, harvesting of macrophytes, fishing, hunting, etc., have consequently led to the depletion of wetland resources also and as such about 85% of our wetlands are in degraded condition. It is, therefore, imperative to preserve these bioaesthetically, economically and ecologically important ecosystems for the benefits of man in future. From the fisheries point of view, the following management measures for the conservation of fish resources vis-a-vis the wetlands are recommended. The rationale upon which such measures are based is that the identity of wetlands as such should be maintained. These include :

(*i*) Manipulation of low water level (0.5-1.0 m) by design.

(*ii*) Maintenance of open water condition.

(*iii*) Ensuring the growth and availability of macrophytes for providing suitable ecological niches for fishes for feeding and breeding purposes.

(*iv*) Protection against predation especially under drought conditions.

(*v*) Banning over-exploitation of fish resources by allowing regularized fishing and enforcing strict legislation.

(*vi*) Banning indiscriminate exploitation of wetlands and their resources by draining, filling, land clearing, burning vegetation, fishing, cattle grazing, harvesting of potential food plants, dumping of domestic refuse, garbage and pollutants and wash off from agricultural fields affecting the food chain severely.

Acknowledgements

This study, a part of the doctoral thesis of the author, was carried out under a research project "Ecology and production of some typical wetlands in Kashmir". The author is thankful to University Grants Commission, New Delhi for providing financial assistance.

References

Bratheke, W.G. and Marchand, M., (1987) Wetlands, the community's wealth. Proc. 3rd conference of Parties to the Ramsar Convention, Regina, Canada. The European Environmental Bureau, Belgium.

De La Cruz, (1973) The role of tidal marshes in the productivity of coastal waters. Assoc. Southeast Biol. Bull. V. 20. pp. 147-56.

Hutchinson, G.E., (1967) A Treatise on Limnology II. Introduction to Biology and Limnoplankton, New York.

Jumppanen, K., (1976) Effects of waste waters on lake ecosystem. Ann. Zool. Fennici. V. 13, pp. 85-138.

Kaul, S., (1982) Community architecture, biomass and production in some typical wetlands of Kashmir. India J. Ecol. V. 9(2), pp. 320-29.

Kaul. V., and Pandit, A.K., (1979) Is Dal lake dying? Science Reporteer V. 16(11), pp. 734-36.

Kaul, V., and Pandit, A.K., (1980) Management of wetland ecosystems as wildlife habitats in Kashmir. In : Proc. International Seminar on Management of Environment (B. Patel ed), Bhabha Atomic Research Centre, Bombay, pp. 31-52.

Kaul, V. and Pandit, A.K., and Fotedar, D.N., (1980) Ecology of snails (gastropod molluscs) in Haigam, a typical wetland of Kashmir. Trop. Ecol. V. 21. pp. 32-46.

Kant, V. and Pandit, A.K., (1901) Benthic Communities as indicators of pollution with reference to wetland ecosystems of Kashmir. In proceedings of who workshop on Biological Indicators and Indices of Environmental, pollution (A.R. Zafar, M.A. Khan, K.R. Khan and Seenayya eds.) Cent. Bd. Water Poll, Cont. Osmania Univ. Hyderabad, pp. 33-55.

Leith, H. and Whittrer, R.H., (1975) Primary Productivity of the Biosphere, Berlin, Springer Verlag.

Lohamn, H. (1908) Unter such ungen zur Fest stellung des Vollstandigen Gehales des Meeres and Plankton. Miss Meeresunters. V. 10.

Maltby, E. (1986) Water-lodged wealth. International Institute for Environment and Development, London and Washington, D.C. p. 200.

Nauwerck, A. (1963) Die Beziehungen Zwischen Zooplankton and Phytoplankton in See Erken. Symbolae Bot. Upsaliensis. V. 17(5), pp. 1-163.

Naulapaa, A. (1966) Eraiden Soumessa esintyven plankter edien tilavunksien Keskria voja, Vesiensuojelutomiston Tiedonantoja, V. 21, pp. 1-26.

Pandit, A.K., (1980) *Biotic factor and food chain structure in some typical wetlands in Kashmir*. Ph.D. Thesis, University of Kashmir, Srinagar.

Pandit, A.K., (1991) Conservation of wildlife resources in wetland ecosystems of Kashmir, India.Proc. Conference, Rennes, J. Environ. Manage, V. 33, pp. 143-64.

Pandit, A.K. and Fotedar, D.N., (1982) Restoring damaged wetlands for wildlife J. Environ. Manage. V. 14, pp. 359-68.

Pandit, A.K., and Kaul, V., (1981) Trophic structure of some typical wetlands. In ; Wetlands, ecology and management, part-ii (B. Gopal, R.E. Turner, R.G. Wetzel and D.F. Whigham, Eds.) Intern. Sci. Publ. & National Institute of Ecology, India, pp. 52-82).

Pandit, A.K., and Qadri, M.Y., (1986) Nutritive values of some aquatic life-forms of Kashmir. Environmental Conservation, V. 13(3), pp. 260-62.

Turner, R.E., (1977) Intertidal vegetation and commercial yields of penaeid shrimp. Trans. Am. Fish. Soc. V. 106, pp. 411-16.

Sawant, P.L. and Magar, N.G. (1961) Studies on frozen fish, II: Some chemical changes occurring during frozen stage. Food Technology V. 15(8), pp. 347-50.

Subla, B.A., (1970) Studies on the functional morphology of the feeding apparatus, feeding habits and the food of nine Kashmir fishes with notes on their ecology. Ph.D. Thesis, Kashmir University, Srinagar.

Water Quality and Periphytic Algal Component of some Macrophytes of Anchar Lake, Kashmir

S.G. Sarwar

Introduction

The lakes of Kashmir are socio-economically and bioaesthetically important aquatic ecosystems harbouring a rich and diverse gene pool. In recent years some of these lakes have shown definite signs of cultural eutrophication due to the increase in the human activities in their catchment area (Zutshi *et al.*, 1980). The lake water, which is a basic component of our natural environment, is being polluted at an alarmingly fast rate. The erosion of lake environment has greatly accelerated during the last two decades causing serious economic loss to the entire society. Enhanced population explosion, rapid rate of encroachments and increased utilization of lake waters for tourist traffic and disposal and dilution of sewage, etc., have not only deteriorated the water quality of these lakes but have also affected biotic flora. Most of the limnological investigations of this region have centred around Dal lake (Mukerjee, 1921, 1926; Kaul and Zutshi, 1966; Kant and Kachroo, 1974; Trisal, 1977; Zutshi and Vass, 1978; Vass and Zutshi, 1979; Trisal and Kaul, 1983), while all other waterbodies, although equally important, have not been given sufficient attention by the limnologists. In view of this detailed limnological investigations of Anchar lake were undertaken for one year from June, 1982 to May, 1983 to provide base-line data on water quality and biotic components.

Material and Methods

Anchar lake (alt. 19583m a.s.l.) is situated at a distance of 14 km to the north-west of Srinagar within the geographical co-ordinates of lat. 34°20′N; long 74°82′74°85′E (Fig. 23.1). The morphometric features of the lake are presented in Table 23.1.

The lake is single basined being connected on its eastern side with Dal lake through an inflow channel "Nalla Amir Khan". A network of channels from the cold water river Sind enter the lake on its western shore forming delta. The water to the lake is also supplied by the springs present in the basin and along the periphery. Spring precipitation also

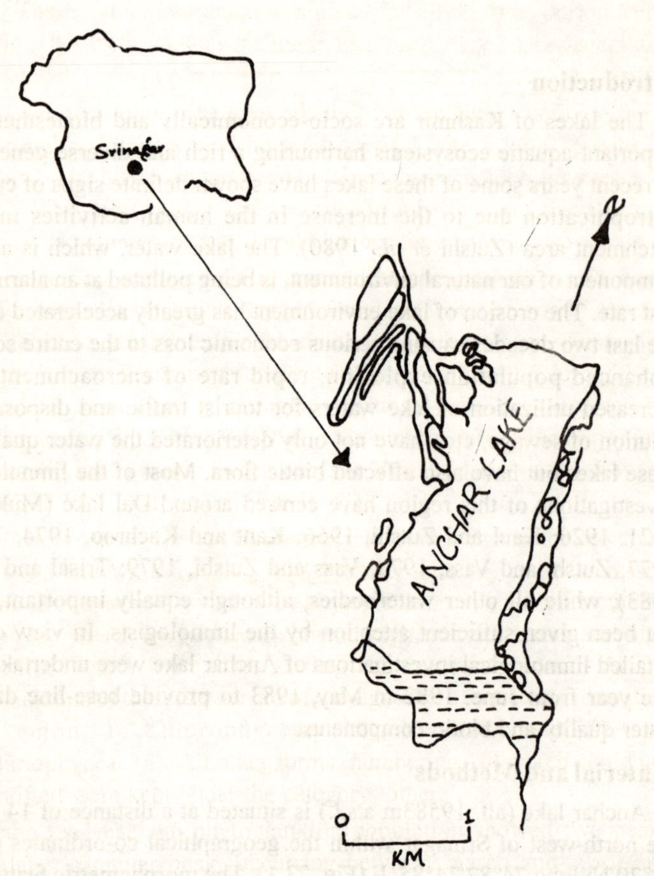

Fig. 23.1 : Map of Jammu & Kashmir showing location of Anchar Lake.

contributes considerable quantity of water. A number of small channels from agricultural fields and human settlements also drain into the lake. Although there is no permanent outflow channel, excess quantities of water leave the lake through an exit in the North-East direction.

Table 23.1 : Morphometric features of Anchar lake

Maximum length	3.55 km
Maximum breadth	0.65 km
Mean breadth	0.60 km
Maximum depth	3.5 m
Mean depth	1.1 m
Surface area	680 ha

The lake has typical rural environment. The catchment comprise long stretches of elevated land on the north-west side which is used for raising different types of vegetables. Agricultural fields surround the lake on the western side. The elevated plateau is partly under paddy cultivation and partly under apple and willow plantations.

The lake supports rich growth of *Myriophyllum spicatum Ceratophyllum demersum, Potamogeton natans, P. crispus, P. lucens, Trapa natans, Salvinia natans, Nymphoides peltata, Hydrocharis dubia, Hydrilla verticellata,Nelumbo nucifera, Phragmgites australis, Nymphae alba, Lemna gibba, L. minor* and *L. trisulca.* According to Zutshi (1975) Anchar lake has greatest diversity of plant communities with twenty associations. Kak (1981) reported 42 macrophytic species from this lake.

Sampling was done at the monthly intervals. The water samples were collected in 1 litre capacity polyethylene bottles and analysed for various parameters adopting the methods of Machereth (1963), APHA (1970) and Goltermann *et al.*, (1978). Grip pH meter No. 121, Systronics conductivity meter type 303 and Systronics Double Cell Colorimeter type 102 were used for determining pH, specific conductance and other chemical characteristics of water samples respectively. Trace metal estimation was carried out on Atomic Absorption Spectrophotometer, model 2900 SP (Pye Unicam Ltd).

Method of Gough and Woelkerling (1976) was adopted for the removal and quantification of periphytic algal flora from *Myriophyllum, Ceratophyllum, Potamogeton natans, Nymphoides, Typha* and *Phragmites.*

Results and Discussion

The ranges and averages for various physico-chemical parameters are depicted in Table 23.2.

The lake is a small sized waterbody with depth ranging from 0.02 to 3.0 m. According to Rawson (1955) lakes with a mean depth of 10 m or less are usually considered as eutrophic. Since the investigated lake has the maximum depth of only 3.5m. it can be considered eutrophic. The light penetration is reduced due to the silt-loaded inflow from the river Sind (Zutshi *et al*., 1980). According to Yoshimura (1933) waters with reduced light penetration are eutrophic.

The minimum water temperature of the lake never fell below 7.5°C and therefore can be classified as warm monomictic or sub-tropical lake (Hutchinson, 1957; Reid and Wood, 1976). The lake waters remained well mixed throughout the investigation period without depicting either summer or winter stratification. Since the maximum depth does not exceed 3.5 m and as such there is no question of stratification as according to Crumrine and Beeton (1975) summer stratification develops only when the depth exceeds 8 m.

The pH values are indicative of the alkaline nature of the waterbody. The seasonal fluctuations in the H-ion concentration were not significant as the waters are well buffered. Zutshi *et al*., (1980) attribute the alkaline nature of the Kashmir lakes to the presence of calcium rich rocks in the catchment areas. Higher silicate content in the rocky beds and sediments also contribute to higher pH due to the production of silicic acid by hydrolysis of the silicates.

The high conductivity values of Anchar indicate high ionic concentration. According to Cumrine and Beeton (1975) lakes of highest conductivity also have greater inputs of ground water. The lower conductance values recorded during spring and summer are attributed to the intake of ions by plants during the growing season while higher values during winter are related to the abundance of nutrients which are released due to the death and decay of macrophytes.

In the investigated lake the range of dissolved oxygen was 3.2-13.6 mg/l ($\bar{x}$: 9.8 mg/l). Maximum DO content was observed during winter months as (*i*) cold waters are known to have more dissolved oxygen than warm waters (Rao, 1955), (*ii*) the rate of decomposition of plant matter was retarded as the water cooled making the micro-organisms less active to respire with the result an appreciable quantity of DO content still remained in the water column, and (*iii*) rich periphytic algal growth and their photosynthetic activity helped to maintain appreciable amount of dissolved oxygen in water.

Table 23.2 : Physico-Chemical features of Anchar lake.

Parameter	*Unit*	*Range*	$\bar{x}$
Water temperature	°C	7.5 – 26.0	17.2
Depth	m	0.02 – 3.0	0.75
Transparency	m	0.02 – 2.0	0.45
Extinction co-efficient	K	1.04 – 28.0	8.22
pH		7.2 – 9.0	7.9
Specific conductivity at 25°C	us	122 – 526	316
Dissolved oxygen	mg/l	3.2 – 13.6	9.8
Oxygen saturation	%	39.9 – 229.7	123.7
Free carbon dioxide	mg/l	0.05 – 26.7	6.9
Magnesium	mg/l	1.9 – 21.8	8.8
Calcium	mg/l	23.2 – 58.0	38.2
Sodium	mg/l	1.3 – 14.4	5.8
Potassium	mg/l	1.0 – 10.6	4.1
Total alkalinity $CaCO_3$	mg/l	58.0 – 275.0	149
Chloride	mg/l	11.0 – 38.0	24.0
Sulphate	mg/l	0.4 – 2.6	0.9
Silicate	mg/l	0.4 – 6.3	1.9
Iron	mg/l	0.06 – 0.4	0.2
Ammonical-nitrogen	mg/l	0.01 – 0.04	0.02
Nitrate-nitrogen	mg/l	0.02 – 0.36	0.09
Ortho phosphorus	mg/l	0.006 – 0.06	0.002
Total phosphorus	"	0.002 – 0.25	0.09

The ionic composition of the lake, computed on an average basis, reads as :

$HCO_3 > CA^{2+} > Cl^- > Mg^{2+} > Na^+ > K^+ > SO_4^{2-}$, which reveals the predominance of calcium and bicarbonate over other ions and in this respect it does not differ much from a "mean" global freshwater composition (Rodhe, 1949). There is slight departure from his generalised Scheme ($Ca^{2+} > Mg^{2+} > Na^+ > K^+$ and $HCO_3 > SO_4^2 > Cl^-$) as Cl^- is dominant to SO_4^{2-} in the investigated lake.

In the investigated lake, total alkalinity was of bicarbonate type. According to the classification of Sawyer and McCarty (1980) Anchar lake water is moderately hard. Following the classifications of Philipose (1959) and Spence (1964) it can be placed in the higher category of nutrient types.

The chloride concentration of the lake may be related to the presence of large amounts of organic matter and sewage contamination. Blum

(1957) also attributes high chloride content to sewage contamination while Thresh *et al.* (1944) relates it to organic pollution of animal origin.

The sulphate content was quite low and is not supposed to play any key role in structuring and shaping of the biotic communities.

Divalent cations-calcium and magnesium exhibited a ratio of 4.3:1 while the monovalent cations-sodium and potassium appeared in the ratio of 1.4:1. Ohle (1934) has laid stress on the importance of calcium in classifying nutrient types of water. He considers lakes with less than 10 mg/l of calcium as poor, with 10-25 mg/l as medium and with more than 25 mg/l as rich. Applying his classification, the investigated lake falls in the 3rd category and as such is polluted. These findings are in agreement with Zutshi (1968) and Vass (1973) who have also categorised lake waters of Kashmir as "calcium rich". Forsyth and McColl (1975) attributed high calcium values to the use of fertilizers which might also be contributing towards total hardness in the present lake as fertilizers are used in its catchment.

Magnesium content is low in comparison to calcium but it is not thought to be limiting factor for the growth of phytoplankton, periphyton or macrophytes. The low magnesium content is possibly due to the uptake of magnesium by the plants in the formation of Chlorophyll-Magnesium Porphyrin metal complex and in enzymatic transformations.

The mean ratios of various salts are presented in Table 23.3. It has been observed that Cl^-/HCO^-_3 ratio was similar to that of Clarke (1924) while other ratios, except ratio for Mg^{2+}/Ca^{2+}, were lower than Clark's ratios for world freshwaters.

The silicate content is low as the Kashmir lakes abound in diatoms (Sarwar 1986, 87, 88, 89) which utilise this mineral. The average iron content of the lake (0.02 mg/l) is in conformity with the values reported by Watzel (1972) for some hard water lakes of U.S.A. Higher iron values may be associated with increase in turbidity and suspended matter as demonstrated by Kothandaraman and Evans (1978).

The relatively low content of nitrate-nitrogen in the investigated lake may be attributed to luxriant macrophytic growth which utilise it during photosynthesis. The low orthophosphate content is due to the formation of an insoluble calcium-phosphate complex. According to Wetzel and Otsuki (1974) such a phenomenon functions as scavenger of some inorganic nutrients and also acts as a removal agent of dissolved matter by absorption. On the basis of total phosphorus values the lake can be placed in the moderate eutrophic category (Strom, 1930), since average

TP values exceed 0.05 mg/l hence there is such a profuse macrophytic growth in the lake.

Table 23.3 : Mean ratios of salts.

Salts	*Anchar lake*	*Clarke's (1924) ratios for World freshwaters*
Cl^-/HCO_3^-	0.16	0.16
Mg^{2+}/Ca^{2+}	0.23	0.17
Ca^{2+}/Cl^{-+}	1.58	3.60
Ca^{2+}/Mg^{2+}	0.31	0.68

Trace-elements like manganese, cobalt, nickle, copper, cadmium and mercury were present below the detection levels. The concentration of lead varied from non-detectable level to 0.31 mg/l ($\overline{x}$: 0.14 mg/l) and that of zinc from undetectable level to 0.51 mg/l ($\overline{x}$: 0.44 mg/l). The low concentration of lead and zinc reflected natural background level. These are within the permissible limits showing thereby that there is no heavy metal pollution.

One hundred fifty-four taxa, representing six algal classes were recorded from the six natural substrates. Bacillariophyceae and Chlorophyceae had the highest representation with 70 and 65 taxa respectively ; while Cyanophyceae (15), Dinophyceae (2), Cryptophyceae (1) and Chrysophyceae (1) were the least represented classes. Fifty-three taxa were common to the investigated substrates.

Class-wise distribution of various taxa on different substrates was as under :

	Typha	*Myrioph-yllum*	*Ceratoph-yllum*	*Potamo-geton*	*Nympho-ides*	*Phragmi-tes*
Bacillariophyceae	47	55	46	52	50	46
Chlorophyceae	31	43	36	34	31	28
Cyanophyceae	10	11	9	11	11	11
Dinophyceae	2	2	1	2	2	2
Chrysophyceae	–	–	1	1	–	–
Cryptophyceae	1	1	–	1	–	1
Total	91	112	93	101	94	88

Myriophyllum supported the maximum number of taxa while the minimum number was recorded on *Phragmites*.

Some of the dominant taxa recorded are : *Achnanthes minutissima, Ampora* spp., *Cocconeis placentula, Cymbella lanceolata, C. ventricosa,*

Fragilaria capucina, F. crotonensis, Synedra spp., *Navicula* spp. and *Oedogonium* sp.

Rare taxa encountered on natural substrates in the investigated lake are : *Caloneis amphisbaena, Cymbella sinuata, Melosira,* sp. and *Neidium dubium* (Bacillariophyceae), *Closterium moniliforme, Pediastrum ovatum, Selenastrum* sp. and *Straurastrum furcigerum* (Chlorophyceae).

On the basis of degree of attachment and proximity to the substratum various algal life forms in the periphyton can be divided into three groups; 1. 'Tightly attached algae' include forms that grow appressed to the substratum (e.g., *Cocconeis, Eunotia*). 2. 'Filamentous and long stalked algae' are also attached to the substratum but extend outward to form a mat (e.g., *Cymbella*) and 3. Motile-planktonic and filamentous long stalked algae.

Community co-efficient of periphytic algal flora on investigated substrates of the lake ranged from 6.21 to 10.04 % ($\overline{x}$: 7.94 %) with the maximum recorded in November, 1982 and the minimum in May, 1983.

Diatoms dominated the periphytic algal component of the lake on all the substrates. Jòrgensen (1957), Stockner and Armstrong (1971) and Ennis (1975) also observed the dominance of diatoms in periphyton assemblages. Numerical superiority of diatoms, both in terms of number of taxa and population density, is indicative of the eutrophic nature of the waterbody. The dominance of diatoms with sub-dominance of blue-green algae is regarded as a characteristic feature of eutrophic waters (Rawson, 1956; Lund, 1962 and Reynolds and Allen, 1968).

On all the substrates the per cent contribution of diatoms was more than 80% (Table 23.4) while the share of other classes was quite low.

The overall periphytic algal population varied from 3×10^4 u/10 mg d. wt. in Aug., 1982 to 23.2×10^4 u/10 mg d.wt. in April, 1983 with an average value of 0.78×10^4 u/10 mg d.wt. In the present study total periphyton population depicted bimodal behaviour with major peak (18.6×10^4 u/10 mg d.wt.) recorded in spring and another peak (13.4×10^4 u/10mg d.wt.) in winter. The peaks are mainly contributed by the diatoms. The abundance of diatom population in cold months is probably due to the fact that they are able to grow under conditions of weak light and low temperature which are lesss suitable for other algae (Lund, 1965; Zafar, 1967).

The summer-autumn decline in the total periphyton population density can be attributed to the grazing effect of zooplankters which abound in Kashmir lakes during this period.

Table 23.4 : Average per cent contribution of major algal classes.

Substrate	*Bacillariophyceae*	*Chlorophyceae*	*Cyanophyceae*
Myriophyllum	91.2	2.8	5.4
Ceratophyllum	85.4	6.8	7.4
Nymphoides	81.4	4.9	11.9
Potamogeton	85.3	3.9	10.6
Phragmites	83.6	5.9	8.3
Typha	95.4	1.1	3.4
On all investigated substrates	91.5	5.0	3.1

The silicate concentration of the investigated waterbody was generally low particularly in the months following peak development of diatoms, which indicates uptake of silicate by diatoms. The low silicate content and the abundance of diatoms in the ecosystems is suggestive of the dependence of diatoms upon the availability of silica. Munawar (1974a, b) and Wetzel (1975) have observed the depletion of silica in ponds and lakes due to diatoms. Thus it can be suggested that the diatoms, which is formed the bulk of periphyton biomass, consumed the silicate content of water (r= -0.51).

However, it appears that the seasonal fluctuations in periphytic population is not governed by a single factor but by the cumulative effect of several physico-chemical factors. Similar observations have also been made by Reid (1961).

Variance in the population density on the investigated natural substrates was quite significant with F value recorded as 4.72 which is higher than the critical value of 2.36 at 5% level of significance.

Bacillariophyceae too showed quite significant variation (F =4.4). In terms of average population density it has been suggested that the highest average values were obtained on *Myriophyllum* followed by *Ceratophyllum* and the lowest on *Phragmites* (Table 23.5).

Table 23.5 : Average population values (u /10 mg. d.wt.) on natural substrates.

Myriophyllum	40,469
Ceratophyllum	28,603
Typha	21,901
Potamogeton	13,991
Nymphoides	13,515
Phragmites	3,310

Myriophyllum and *Ceratophyllum* supported large population density as they offered greater surface area due to the dissected nature of their leaves. Harrod and Hall (1962) and Cattaneo and Kalff (1979) also recorded similar observations.

Periphyton population showed positive correlation with pH (r = 0.017), chloride (r = 0.109), sodium (r = 0.057), sulphate (r = 0.0216 dissolved oxygen, orthophosphorus (r = 0.347) and total phosphorus (r = 0.162). It depicted weak negative correlation with water temperature (r = -0.371) nitrate-nitrogen (r = -0.260), ammonical–nitrogen (r = -0.430), specific conductivity (r = -0.049), iron (r= -0.008), potassium (r = -0.085), magnesium (r = -0.215) and calcium (r = -0.41); and a significant inverse correlation with silicate (r = -0.521) and total phosphorus (r = -0.538).

In conclusion, it may be said that the investigated lake shows signs of accelerated rate of cultural eutrophication due to the stresses and strains of the biotic interferences. This important waterbody is shrinking and getting deteriorated at an alarmingly high rate due to human greed (accelerated land use along the catchment, dumping of domestic refuse, garbage and sewage, and illegal encroachments, etc.) and it is losing its character and identity as it is rapidly becoming silted up. An integrated and multidisciplinary approach is needed to conserve this lake from further degradation. The various conservation measures to be adopted should include : (*a*) reduction of the nutrients entering the lake by catchment improvement and control by diversion of sewage and inflows from the agricultural fields and orchards, etc. (*b*) harvestation and removal of the weeds to improve the nutrient balance of the lake. (*c*) afforestation of the catchment area. (*d*) ban on illegal encroachments (*e*) regulation and curtailment of the silt nutrient load flowing into the lake. (*f*) control of the water–level fluctuations of the lake.

Acknowledgements

I am grateful to the Principal, S.P. College, Srinagar and Director, C.O.R.D., University of Kashmir for providing necessary facilities for these investigations, and to U.G.C. for providing financial assistance during the course of these studies.

References

American Public Health Association, (1970) Standard methods for the examination of water, sewage and industrial wastes. APHA, New York.

Blum, J.L., (1957) An ecological study of the algae of the saline river, Michigan. Hydrobiologia, V. 9(4), pp. 361-408.

Cattaneo, A. and Kalff, J., (1979) Primary production of algae growing on natural and artificial plants : A study of interactions between epiphytes and their substrate. Limnol. Oceanogr. V. 24, pp. 1031-37.

Clarke, F.W., (1924) The Data of Geochemistry, 5th ed. Bull U.S. Geol. Surv., 770, pp. 841.

Crumrine, J.P. and Beeton, A.M., (1975) Limnology of lakes of the Sylvania Recreation Area, Ottawa National Forest Special Report No. 24. Centre for Great lake Studies, the University of Wisconsin, Milwakee, U.S.A., pp. 34.

Ennis, G.L., (1975) Distribution and abundance of benthic algae phosphorus gradients in Kootenay Lake, British Columbia. *Verh. Int. Ver. Limnol.* V. 19, PP. 562-70.

Forsyth, D.J. and McColl, R.H.S., (1975) Limnology of lake Nagahema, North Island, New Zealand, N.Z.J. *Mar. Freshwat. Res.* V. 9(3), pp. 311-22.

Golternan, H.L., Clymo, R.S. and Ohnstrad, M.A.M., (1978) Methods for physical and chemical Analysis of Freshwaters, IBP Handbook No. 8.

Grough, S.B. and Woelkerling, W.J., (1976) On the removal and quantification of algal aufwuchs from macrophyte hosts, Hydrobiologia. V. 48, pp. 203-07.

Harrod. J.J. and Hall, R.E., (1962) A method for determining the surface area of various aquatic plants. Hydrobiologia, V. 20, pp. 173-78.

Hutchinson, G.E., (1975) A Treatise on Limnology 1. Geography, Physics and Chemistry. New York, John Wiley and Sons, Inc.

Jorgensen, E.G., (1957) Diatom periodicity and silicon assimilation. Dansk Botl. Arkiv. V. 18(1), pp. 54.

Kak, A.M., (1981) Floristic composition, structure and ecology of some high altitude lakes of Kashmir Himalayas. proc. Abs. Silver Jubilee Symposium of Tropical Ecology "Ecology and Resource Management in the Tropics. Ed. R.S. Ambasht and H.N. Pandey, Varanasi, pp. 106-08.

Kant, S. and Kachroo, P., (1974) Limnological studies in Kashmir lakes. IV. Seasonal dynamics of phytoplankton in the Dal and Nagin. Proc. Indian Nat. Sci. Acad., India, V. 39(B) 5, pp. 632-45.

Kaul, V., and Zutshi, D.P., (1966) Some ecological considerations of floating islands in Srinagar lakes. Proc. Indian Nat. Acad. Sci, India, V. 36, pp. 273-81.

Kothandaraman, V. and Evans, R.L., (1978) Physical and Chemical water quality characteristics of the Fox Chain of lakes. Trans. III. Stat. Acad. Sci. V. 71(3), pp. 239-59. Rep Series No. 462 III. *Strat. Wat. Surv.*

Lund, J.W.G., (1962) Phytoplankton from some lakes in Northern Sarkatchewan and from Great Slave Lake. Can. J. Bot. V. 40, pp. 1499-514.

Lund, J.W.G., (1965) The ecology of freshwater phytoplankton. Biol. Rev. V. 40, pp. 231-93.

Mackereth, F.J.H., (1963) Water Analysis for Limnologists. Freshwater Biol. Assoc. V. 21, pp. 1-70.

Mukherjee, S.K. (1921) The Dal lake (Kashmir) – A study in biotic succession. *Proc. 8th Indian Sci. Congr.* pp. 185 (Abstract).

Mukerjee, S.K., (1926) Aquatic and marsh vegetation of the Dal lake of Kashmir. *Proc. Linn. Soc.,* London. pp. 55 (Abstract).

Munawar, M., (1974a) Limnological studies on freshwater ponds of Hyderabad, India.III. The Biocenose. Seasonal abundance of unicellular and colonial phyplankton in polluted and unpolluted environments. Hydrobiologia. V. 45(1), pp. 13-27.

Munawar, M., (1947b) Limnological studies on freshwater ponds of Hyderabad, India. IV. The Biocenose. Seasonal abundance of unicellular and colonial phytoplankton in polluted and unpolluted environments. Hydrobiologia. V. 45(1), pp. 1-32.

Ohle, W., (1934) Chemische and Physikalische Untersuchungen norddeutscher Seen. *Arch. Hydrobiol.* V. 26, pp. 386-464 and 584-658.

Philipose. M.T., (1959) Freshwater phytoplankton of inland fisheries. Proc. Symp. Algol., I.C.A.R., New Delhi, pp. 271-91.

Rao, C.B., (1955) On the distribution of algae in a group of six small ponds. II. Algal periodicity *J. Ecol.* V. 43, pp. 291-308.

Rawson. D.S., (1955) Morphomentry as a dominant factor in the productivity of large lakes *Verh. Int. Ver. Limnol.* V. 12, pp. 164-75.

Rawson, D.S., (!956) Algal indicators of trophic lake types, Limnol. Oceanogr. V. 1, pp. 18-25.

Reid, G.K., and Wood, R.D., (1976) Ecology of inland waters and Estuaries. D. Van. Nostrand Co., New York, pp. 485.

Reynolds, C.S. and Allen, S.E., (1968) Changes in the phytoplankton of

Oak Mere, following the introduction of Base-rich water. Br. Phycol. Bull. V. 3(3), pp. 451-62

Rodhe, V., (1949) The ionic composition of lake waters. *Verh. Int. Ver. Limnol.* V. 10, pp. 377-86.

Sarwar, S.G., (1986) Species composition and seasonal variations of periphyton on Ceratophyllum demersum Linn. in Waskur lake, Kashmir. Geobios New Reports. V. 6, pp. 114-18.

Sarwar, S.G. and Zutshi, D.P., (1987) Studies on periphyton population of Himalayan lakes 1. Species composition and community structure on natural and artificial substrates. Proc. Indian Nat. Sci. Acad. V. 53, pp. 239-43.

Sarwar, S.G., and Zutshi, D.P., (1988) Species distribution and community structure of periphytic algae on artificial substrate. Trop. Ecol. V. 29(2), pp. 116-20.

Sarwar, S.G., and Zutshi, D.P., (1989) Species composition and population density of periphytic algal flora of two macrophytes of Kashmir. Indian J. Ecol. V. 16(1), pp. 68-71.

Sawyer, C.N. and McCarty, P.L., (1980) Chemistry for Sanitary and alkali soils. Agri. Handbook 60, U.S. Dept. of Agric. Washington D.C., pp. 100. In : David Keith Todd (ed) : Groundwater Hydrology. John Wiley & Sons, Inc.

Spence, D.H.N., (1964) The macrophytic vegetation of lochs, swamps and associated fens. pp. 306-425. In : J.H. Burnett (ed.) : the vegetation of Scotland. Oliver and Boyd, Edinburgh.

Som, K.M., (1930) Limnological observations on Norwegian lakes. Arch. Hydrobiol. V. 21, pp. 97-124.

Stockner, J.G., and Armstrong, F.A.J., (1971) Periphyton of the Experimental Lakes Area (ELA), Northwestern Ontario, J. Fish. Res. Board Can. V. 28, pp. 215-29.

Thresh, J.C. Suckling, E.V. and Beale, J.E., (1944) The examination of water and water supplies. 6th Edition (E.W. Taylor, ed.).

Trisal, C.L., (1977) Studies on primary production in some Kashmir lakes. Ph. D. Thesis, Kashmir University (Unpublished).

Trisal, C.L. and Kaul, S., (1983) Sediment composition, mud-water interchanges and the role of macrophytes in Dal lake, Kashmir. Int. Revue Ges. Hydrobiol, V. 68, pp. 671-82.

Vass, K.K., (1973) Studies on hydrobiology and primary productivity of some Kashmir lakes. Ph. D. Thesis, Kashmir University (Unpublished).

Vass, K.K. and Zutshi, D.P., (1979) Limnological studies on Dal lake. 1. Morphometry and physical features. J. Indl. Fish. Soc. India, V. 2(1), pp. 12-21.

Wetzel, R.B., (1972) The role of carbon in hard-water marl lakes. In : G. E. Likens, ed. Nutrients and Eutrophication : The Limiting-Nutrient Controversy. Sp. Sym. Amer. Soc. Limnol. Oceanogr. V. 1, pp. 84-91.

Wetzel. R.G., (1975) Limnology. W.R. Saunders Co., Philadelphia, pp. 743. Wetzel, R.G. and Otsuki, A., (1974) Allochthonus organic carbon of a marl lake. *Arch. Hydrobiol.* V. 73, pp. 31-56.

Yoshimura, S., (1933) Rapid eutrophication within recent years of lake Haruna, Gunman, Japan, *J. Geol. Geog.* V. 2, pp. 31-41.

Zafar, A.R., (1967) On the ecology of certain fish ponds of Hyderabad, India, III. The periodicity. Hydrobiologia. V. 24(4), pp. 556-66.

Zutshi, D.P., (1968) Ecology of some Kashmir lakes. Ph.D. Thesis, Jammu and Kashmir University (Unpublished).

Zutshi, D.P., (1975) Associations of macrophytic vegetation in Kashmir lakes. *Vegetatio,* V. 30, pp. 61-65.

Zutshi, D.P. and Vass, K.K., (1978) Limnological studies on Dal lake. II. Chemical features. *Indian J. Ecol.* V. 5(1), pp. 90-97.

Zutshi, D.P. Subla, B.A., Khan, M.A., and Wanganeo, A, (1980) Comparative limnology of nine lakes of Jammu Kashmir Himalayas. Hydrobiologia, V. 72 (1-2). pp. 101-12.